By the Author of

Raising the Pentagon

The
Dynamics
of Flight

A woman approaching her 30th birthday takes off to confront life and learns about death

An Adventure on the Road

by

Robin L Stratton

MOCKINGBIRD SQUARE

I would like to gratefully acknowledge the help of several people, beginning with my parents, who continue to encourage me, and my brothers, Smitty and Jay. I also want to thank Wayne R. Petersen, a field ornithologist for the Audubon Society, who offered much appreciated (and much needed) advice about bird watching. Special thanks go to Jim, for inspiration. Most of all, I want to extend thanks to all the bookstore managers who helped me so much with my last book. **You know who you are.**

The Dynamics of Flight. © 1991
by Robin L Stratton

Illustrations and cover by the author

MOCKINGBIRD SQUARE
P.O. Box 3
Wilmington, MA 01887

√√√√√

Printed in the United States of America

MOCKINGBIRD SQUARE

Publisher

This book is dedicated to everyone who is about to turn 30, everyone who has already turned 30, and everyone who turned 30 a long time ago and never quite got over it.

CHAPTER ONE

Alex and I made love quietly but not without passion. Immediately afterwards he went to kiss my cheek, but hit my ear instead, making it ring with painful unexpectedness. And then only a second later he was sound asleep. As I listened to his deep, slow breathing, it occurred to me that he hadn't even taken the time to tell me he loved me. That kind of pissed me off. Besides. He had a cold, and oxygen was having a tough time making it through his nostrils. It wasn't quite snoring, but it was damn close. I wanted desperately to wake him up and make him blow his nose. I even pictured me doing that, and I imagined what his reaction would be. He'd wake with a start, and he'd stare at me, then he'd say, "What." I'd feel bad about waking him, so I'd probably feign innocence. I'd probably say, "What" too. He'd say, "I thought you said something," and I'd say, "No. Go back to sleep." And he would.

After a minute, I sighed quietly. I thought about how quickly he could fall asleep, and I was envious and amazed. For as long as I can remember, I've never been able to fall asleep less than an hour after I've gone to bed. But Alex was one of those incredibly lucky people who lose consciousness the moment they stop moving. There was a time, back when we first started living together, that it really bothered me, because I'd read in about half a dozen magazines that women have a right to expect a little sweet talk after love making; but apparently Alex hadn't read the same magazines I had. I doubted that *BusinessWeek* and *Newsweek* explored issues like romance after intimacy. Turning my head to the wall, I watched a cobweb hula in the lamplight. I'd stared at that same cobweb every night for almost a week. It was probably time to do something about it.

The radio was still on. One of those sugary Michael Jackson ballads that I secretly liked. I listened it out to the end, then reached over to shut it off. For the briefest moment, I wondered if the sudden cessation of music would wake Alex up. But then right away I scoffed quietly. No way. He was incredible, he could sleep through anything.

Or could he? All at once I was childishly inspired to test his powers. So I made a real racket, heaving a loud, theatrical sigh, snapping the sheet over my shoulders and moving around so much that the water in our waterbed sloshed. But he just waited out the noise, then snuggled up against me. There'd been only the slightest intermission, and then the snoring resumed. Only now it was right on my chest. Stroking his thick, dark hair more out of habit than actual affection, I wanted to cry. I wanted to scream, "How can you sleep? I'm wide awake! I'm lonely and depressed, and you don't even know it!"

But of course I didn't. I just lay there, watching the cobweb and stroking his hair and listening to him sleep.

Then I thought that maybe sometime soon I should quit whining about him. Alex was the sweetest guy I'd ever met, next to my father. He was generous and thoughtful and unselfish. He always sent flowers on my birthday and our anniversaries (which he remembered better than I did, including the date of the very first time we made love) and he even bought me cards for no reason, those "Just thinking of you" ones. Best of all, he didn't seem to mind that I was about to become an old woman.

"Turning thirty is nothing, Lark. Remember when I turned thirty last year? Didn't bother me a bit."
"It's different for a man, though," I grumbled.
"Why."
"I don't know. It just is."
"Just a state of mind. If concepts like 'years' and 'birthdays' had never been invented, you wouldn't have any idea how old you were, and you wouldn't be upset."
"But those concepts were invented, and I know exactly how old I am! Almost thirty!"
"Well so what? Do you think your life is suddenly going to grind to a halt just because of some date on the calendar?"
I nodded sullenly. That date was only a week away.
"Look," he said, "you're beautiful, intelligent, and you have a man who loves you and would do anything in the world for you. What more could you possibly ask for."
"Would you really do anything in the world for me?"
"You bet. Just name it."
"Make me twenty-five again."

Abruptly, my eyes grew damp with self-pity. A couple of tears dribbled down the sides of my eyes, and crept into my ears, which tickled and irritated me, but I didn't bother to do anything about it. I just lay there, crying a little, and hoping that Alex would wake up and insist on knowing what was wrong. But of course he didn't.

"Shit," Alex said when the alarm went off. With a groan, he reached across me and pressed the snooze button. I could detect the faintest fragrance of his sweat, and it made me feel warm and comfortable.

"What's the matter," I said.

"I have to pick up some flowers for my mother today."

"Oh yeah. For her birthday."

"I don't know when I'll have time," he said, sighing deeply. I didn't say anything right away. I knew he was hoping I'd offer to get some. After all, it was Saturday. I had the day off. Alex sighed again. Tragically inconvenienced.

"Would you like me to . . ."

"Would you?" he interrupted gratefully. "I would appreciate that so much." His breath was hot and smelled complacently of sleep, which was not unpleasant. As he nuzzled my neck, I heard him say he loved me. I said I loved him, too, but already I was dreading the day. I hate to shop. I hate those saleswomen who greet me so cheerfully, with bright smiles, asking if they can help me. I say, "No," and they say, "Well, what are you looking for?" and before I know it, they're helping me. In dress shops they select an outfit for me, and hover outside the dressing room while I try it on. Then they poke their heads inside and say, "Ooh, that looks *yummy* on you!" No matter what I do, I can't get them to leave me alone. Unless, of course, I need their help. Then they're nowhere to be found.

Sure enough, the moment I entered the flower shop, a elderly woman with bluish hair that had long ago outgrown its style flashed me a brilliant retail grin. I saw that she wore very round glasses which reminded me first of an owl, then of John Lennon. They rode so low on her nose that I wanted to reach over and push them up. Her breasts were big and pointy, like the kind you see in those movies from the 50s. I said, "Hi" and she said, "Hi" and asked if needed help. I said, "No" and she asked if I was looking for anything in particular.

"Flowers," I said.

"Ah, yes, *flowers*," she said so merrily that I was tempted to stick my tongue out at her. Instead I just looked away. Some pots of very vivid fuchsia — a favorite of the Ruby-throated Hummingbird — attracted my attention.

"Aren't those *lovely*," cooed the saleswoman, having noticed my interest.

I didn't answer. Would Mrs. Hawkins like a plant instead of flowers? Alex always sent flowers. Would she have a stroke and die if a plant arrived instead?

"I'll take one," I said.

It wasn't that I hated Alex's mother. She hated me. She resented me for moving in with her precious son without going through the trouble of marrying him. Well for crying out loud, it wasn't as if my mother was particularly thrilled about the situation, either.

"I can't understand why you don't just marry him. You love him, don't you?"

"Yes, Mother, I love him. But marriage is . . . I don't know. It's such a big step. I would hate to make a mistake."

"Lark, you've been living with Alex for two years. If you don't know by now . . ."

"Mother, what difference does it make if Alex and I are married?"

"It makes a difference to me," she pouted.

I wanted to remind her that we were talking about my life, not hers, but I didn't. I said quietly, as if I didn't want her to hear me, but really I did, "I've seen what marriage can do to a person."

And she heard me, alright. Loud and clear.

"Don't start that again," she said.

"Start what," I said, even though I knew.

"You're talking about your father, aren't you."

I couldn't deny it. I was.

"You know, Lark, I think your father would have wanted you to marry Alex."

"Mother. Stop it." *I covered my ears, the way I always did when she talked about him. I even began to sing to drown out her words. She stopped speaking, helpless in the face of such flaming immaturity.*

"Plenty of light, water regularly," chirped the saleswoman who'd selected a fuchsia for me while I was lost in thought. Meekly, I followed her to the register. I gave her Alex's mother's address, but when I signed the card, I momentarily forgot the circumstances, and signed my name first. Oops. Now Mrs. Hawkins would know I picked out the plant, not Alex. That meant she would hate it. She might even try to kill it. I considered writing another card, but the saleswoman had already slipped it into a tiny yellow envelope. So I just paid and left.

Outside the sun was hot and squinty bright. I automatically looked up to the sky. A red flash caught my eye, and came to land on a telephone wire above me. My first guess was that it was a Purple Finch, but then I saw that it was too slim and more brick colored than raspberry, so that meant it was probably a House Finch. It's a pretty bird, with a bright rosy breast and brown stripes on its belly. For several seconds I stood there, admiring it. But before too long, it flew off. I watched it go. After nearly thirty years, it still fascinated me to see a living creature go cutting effortlessly through the air like that. When you stop to think about it, the dynamics of flight are pretty amazing. Imagine being able to just leap into the sky and disappear? Without having to tell anyone where you were going? Just soaring off and not worrying about . . .

"Hey, Lark! Lark!"

Startled, I returned my attention to my surroundings. After squinting into the sun, it took a moment to focus on the figure I saw heading toward me. He was a good-looking guy, well built and vigorously tanned, strolling up to me with a big grin.

"Bird watching, as usual," he greeted me.

"Hi, Nick," I said with a shrug to acknowledge that my ornithological inclinations were well known among, and of no particular interest to, my friends.

"You look great."

I glanced down at my Audubon tee shirt knotted high above the waist, and the denim miniskirt I wore so often that Alex always teased, "Hey . . . is that skirt new? I've never seen it before," and opted to ignore Nick's compliment, which struck me as being somewhat obligatory.

"What are you up to today," I asked him.

"Lunch. Care to join me?"

I hesitated.

"On me," he added.

It was tempting. I couldn't remember the last time I'd been taken out to lunch. Alex, who was in retail, worked every Saturday, and on Sundays he just wanted to stay home and relax with *The Boston Globe*. I knew he wouldn't mind if I let Nick buy me lunch. That was one of the most exasperating things about him — he was absolutely lacking in anything that even remotely resembled jealousy. He always said he trusted me too much to ever question my behavior. In a way it was kind of nice. But in another way, it was an overwhelming responsibility. I always had the feeling that I could sleep with someone else, and Alex would merely shrug in his gentle, solemn way, and say that he understood. I, on the other hand, flew into a really impressive rage if I thought he was even considering glancing at another woman. The first time I visited him at his store the summer we met, I found it swarming with young girls clad only in shorts and bikini tops. "No wonder you don't mind working on Saturdays!" I roared as soon as he got home. But he only regarded me blankly and said, "Even if you weren't the most beautiful, sexy woman in the world — which you are — I still wouldn't even notice anyone else. I'm completely in love with you." Well for crying out loud, what could I say to that? So I said I was sorry. And he said, "Why are you apologizing? You haven't done anything wrong." Swear to God, sometimes it was like living with Gandhi or something.

"Well?" Nick interrupted my thoughts, bringing me abruptly back into the present. He was watching me speculatively.

"Okay," I said, "let's go."

CHAPTER TWO

We went to an Italian place, Luigi's. Inside it was unbelievably humid, and the odor of foody sweat permeated everything. Within seconds, my Audubon tee shirt grew damp and clingy. Irritably I tugged at my knot while I consulted the menu above the register. As usual, my eyes flew instantly to the sign I've always hated, "Italian food at it's finest." I always wanted to point out to the management that the apostrophe in "its" didn't belong. But of course I never did.

"What will you have," Nick asked, and I saw he was watching the increasing amount of my stomach exposed each time I pulled on the knot. I didn't answer right away. I was thinking that I really shouldn't be eating a big meal in the middle of the day. A big supper was okay, but eating a greasy, Italian lunch would probably leave me feeling guilty for the rest of the day. Not that I'm fat or anything. But when I moved in with Alex I put on fifteen pounds that it wouldn't exactly kill me to lose.

I heard Nick say, "Well?" and I returned my attention to the menu. I'd begun to consider the Tastee Slice o' Pizza with a decisive eye.

"Well, what looks good to you," I said. He didn't answer, and when I looked at him, I saw that he was giving me a sexy smile as if to say he'd like to take a bite out of *me*. Unimpressed, I smirked and stopped tugging at my knot.

"Thinking about the Tastee Slice o' Pizza," he said after he'd fail to slay me. I told him that was what I wanted, too, and we agreed that a couple of slices o' mushroom would really hit the spot.

We ordered, and a big, sweaty Italian guy, presumably Luigi himself, nodded, wiped his dripping forehead with the back of one wrist, and said something neither of us understood but agreed to anyway. I was watching the beads of sweat carefully to make sure none of them rolled off and landed on the pizza. I didn't know what I would have done if I'd seen one; I doubted I would say anything about it. After a couple of minutes the suspense and indecision began to spoil my appetite, so I told Nick I'd get us a booth.

I seated myself at the only empty table, which was chipped at one corner. The seats were still warm and moist from the previous patrons. A moment later, Nick joined me, and as he set down the tray, he made a polite inquiry as to Alex's well being.

"He's fine."

"You tired of him yet?"

I didn't answer right away. Before I met Alex, Nick had asked me out a couple of times, and once he'd even said, "Let's cut through the bullshit. You want to sleep with me?" I'd cut through the bullshit, too, and said No. I remembered that when I told him I was moving in with Alex, he seemed disappointed. I wondered if he was secretly in love with me, or if he saw all women as vaginas to be vanquished, whether he actually wanted them or not.

16

Finally responding, I felt myself shrug. I didn't mean to. I meant to shake my head emphatically and declare, "Never!" But it came out as a shrug. I don't know why. Nick's eyebrows rose.

"I see," he said.

"You see what," I said.

"Oh, nothing," he smiled, and I felt uncomfortable and guilty all of a sudden. Nick raised his cup of Pepsi, and waited as I followed suit. "To true love," he warbled innocently, and our cups met with a cardboardy TAP. He was staring at me, but I couldn't meet his eyes. I should say something, I thought, but I didn't. Instead, I heard myself sigh.

"Lark, I can tell something is wrong between you and Alex. What is it?" Nick wiped some oil off his fingers, then, with that hand, touched my wrist sympathetically.

"Nothing's wrong," I snapped unhappily.

Nick retracted his hand. We ate our pizza in silence for a little while.

"Nothing's wrong," I repeated eventually, "It's just that, I don't know."

"He been treating you okay?"

"You kidding? Alex treats me like a queen! But, well, our relationship isn't that exciting anymore. You know, nothing is wrong, but something's not quite right. I feel like I'm not as happy as I could be. I feel like I'm missing out on something."

"Missing out on what?"

"Missing out on . . . I don't know . . . my life, maybe. I'm about to turn thirty, Nick, and I haven't *done* anything. I feel like I should be *doing* something."

"Like what." Nick had taken an enormous bite of his slice o' pizza, and as I watched, oil dripped down his forearm, two mushrooms slid off onto his plate, and his napkin, snatched up

17

by a muggy gust of air, soared off — all at once. It was a messy, busy interlude for him, and I waited as he pulled himself together again before I continued.

"Well, I could be travelling, or meeting new and exciting people, or embarking on a new career, or sighting a Painted Bunting, or something."

"What the hell is a Painted Bunting?"

"You don't know?"

"Should I?"

"The Painted Bunting is one of the most colorful birds in North America. My father and I always wanted to see one."

"Why haven't you."

"They're not around here. You have to go south. To North Carolina."

"Oh."

I heaved a great big sigh, and went on, "I just have this hideous feeling that time is passing me by. Like my life has no excitement."

"Excitement?" Nick echoed with a greasy smile. "I'll show you excitement."

I frowned. That was the problem with Nick — life, to him, was one big sexual conquest. He once told me that the most sophisticated thing in the world was the desire to get laid.

"You know what I mean. I love Alex and everything . . . but something isn't right. I've been living with him for two years, and it's been really great . . . but for some reason, I'm just not happy. I know I should be. I'm just not. And I don't know why. I feel like I haven't really enjoyed myself in the longest time. I can't remember the last time I laughed until tears came out of my eyes and my sides ached. I used to laugh like that, you know, years ago. I don't know. Maybe I'm too idealistic. But I need more passion in my life."

"I'll show you passion," Nick offered. I was really getting tired of him, and occupied myself with my pizza. He said, "Look. You're just at a point in your life where you need change. Your unhappiness has nothing to do with Alex. What you need to do, Lark, is start taking some risks."

"Start taking some risks," I repeated blandly. Luigi made a great pizza, but it was difficult to eat it and maintain some semblance of ladylike daintiness. I debated whether or not to use my fork. I've always hated eating pizza with a fork. It's so proper. It's like those people who insist on using chopsticks whenever they eat Chinese food. Glancing at Nick, I saw that he'd opted to use his. I buckled under the pressure and picked up mine, too.

"Yeah. You need to shake up your life so that you'll be glad to have it return to normal," he said. He was cutting up his pizza into tidy squares and dispatching them neatly as he spoke. "You need to do shit you wouldn't usually do. You need to get your head out of the clouds and experience reality. Filth. Decadence."

"I do?"

"Uh huh," he nodded. I dragged a napkin — my third since we'd sat down — across my lips and chin, and pondered Nick's advice. Did I want to do that? Take risks? Shake up my life? Experience reality?

"There must be some other way," I said.

"Nope. Trust me on this. You've got to spread your wings."

His ornithological cliché made me smile. I didn't say anything as we finished our pizza. He'd paid, so I thanked him as we prepared to part at the door.

"Think about what I said," he urged. I assured him I would. There followed a moment, during which I could tell he was trying to decide if he could get away with kissing me on the lips or not, and finally he just gave me a hasty hug, which I loosely returned.

"Well, see you," I said.

"Hey wait, I got something for you," he said, plunging his hand into the front pocket of his shorts. Nick worked for a novelty gift manufacturing company, and had a bizarre array of useless items on his person at any given time, which he liked to give out to friends. I had a drawer full of walking, clattering teeth, bogus dog shit, and plastic ice cubes with flies in them. As I waited, he pulled out a bright yellow button with red letters that said, "Life sucks then you die."

"That is absolutely the most dismal and cynical thing I have ever seen in my whole life," I said, taking it and dropping it into my purse.

"Maybe so," he said, "but they're selling like the proverbial hot cakes."

CHAPTER THREE

From Luigi's, I went to my office at the Pack Your Bags Travel Agency. We weren't open on Saturdays, but I'd left my current book, *1001 Facts About Birds*, in my desk, and I wanted to read it over the weekend.

Even before I let myself in, I smelled cigarette smoke. That meant Pat was there. Probably catching up on her filing. She always let it go until several unstable stacks of folders covered her desk, and every five minutes or so I'd hear her say, "Now where did I put that . . ."

I went in, and sure enough, there she was, sorting through a pile of papers. A cigarette hung from her lips and it made me think of a thick, white caterpillar emerging from a hole in a tree. It fascinated me the way she didn't know how horrible she looked when she smoked. And she stunk, too. In an effort to overpower the smell of smoke, she doused herself each morning in perfume, not the real expensive brand, but the budget stuff that was supposed to smell just the same. Sometimes the combination was enough to make me feel like I wanted to puke. And her teeth were stained. I always wanted to say, "How can you do that to yourself? How can you willingly inhale so much nicotine? Don't you know they use that in insecticides, for crying out loud?" but of course I never did.

She looked up as I entered, and gave me a grey smile. At the end of her cigarette was an inch and a half of ash. I could tell it was going to fall off any second and land on the carpet, and I'd have to vacuum it up on Monday.

"Hello, Cherub!" she greeted me. She always called me that, and I always hated it, but for some reason, I never asked her to stop. "Come in to help me file?"

"No," I said, "I need to pick up my book."

Another thing that bothered me out about Pat was that she had these real long fingernails that she always painted bright red, even when her outfit was pink. They were those kind you buy and glue on and polish. I never understood how she could type, they were so long. As I watched her, she went over to the filing cabinet, opened a drawer, leafed through the contents, dropped the folder in, and shut the drawer with a vigorous swing of her hip. The drawer went SQUEAK then SLAM! Turning to me, she took a long drag on her cigarette. The ash tumbled to the floor.

"What book," she said, tapping her talons on the file. Then, before I could answer, she went on, "You know what? Bradley Crenshaw III was in here yesterday after you left. He was asking about you."

I'd crouched at my desk to retrieve my book from the bottom drawer. Pat's news sent a hot flush through me. Grateful that my face was obscured, I said, "Oh, really?" very casually.

But Pat wasn't fooled. She said, "Yes, *really*," and came over to stand near me so that all of a sudden I was staring at her ankles. She was a heavy woman and her ankles were thick. Her wide feet, squeezed into black flats, reminded me of Fred

Flintstone's feet.

"So what did you tell him."

"I'd told him you'd gone home. He said that was too bad. And he asked if you'd be in on Monday."

"Did you tell him I would be?"

"Of course I did, Cherub! I told that gorgeous hunk of ass that you would take care of him *personally.*"

"Pat, I really wish you wouldn't tell him things like that. You know I can't . . . I mean, I have Alex and stuff." I rummaged through the drawer, found my book, pulled it out, and stood. Pat, who was several inches shorter than me, squinted as she looked up, and a thin line of smoke found its way in my nostrils. I tried to step back a little, but my chair was in the way. So I turned my head to the side and wondered what I would do if I had a whole bunch of unattractive habits like her. I wondered what she would do if she ever saw a video tape of herself, and had to look at all the horrible things she did. Once a friend of Alex's filmed me while I was talking, and I couldn't believe how I kind of bobbed my head when I spoke, and shut my eyes for prolonged periods of time when I was trying to think.

"Why, Lark," she said in exaggerated surprise, "I didn't even realize that you and Alex were still together! I mean, you *never* mention him."

I wanted to be pissed at her insinuation, but I didn't have the right. It was true — poor Alex's name just never seemed to come up in conversation anymore.

"Be honest," she said, and with one of her long, terrible false nails, she tapped the air in front of my face, "if Bradley Crenshaw III asked you out for caviar and Perrier, would you say, 'No, thank you, I have Alex and stuff?' I think not."

"He's not going to ask me out. Stop that." She was

bugging the living shit out of me. Bradley Crenshaw III was a dark, sophisticated, intense looking guy several years older than me who'd asked me to arrange several exotic trips for him. He always went alone. Rumor had it he was incredibly rich, and was traveling around the world in search of a wife. Pat and my other coworker, Alice, teased me about the way he always asked for me. "He's got his heart on you," Alice always told me, but the way she said it, it sounded like "He's got his *hard on* you." Then she and Pat would giggle. It got really annoying when he came in. They'd titter and sneak glances and make faces. It was like being back in junior high.

"Well I think he *is* going to ask you out. I think he'll ask you to marry him, and pretty soon Alice and I will be arranging trips for the two of you. You'll be Mrs. Bradley Crenshaw III."

"Stop that," I said again, but by now I was smiling a little. Imagine being so rich you could just drop everything and go to any place in the whole world! I've always loved to travel, but for some reason, I've hardly even left my home state. When I was younger, my father and I used to take long drives. But that was years and years ago. Since then, I haven't strayed much from the range of my office, my apartment, my parents' house, and Alex's parents' house. "Free as a bird," I murmured, not even aware I'd spoken aloud.

"Not free," Pat corrected. "You'll have to work your ass off. What'll you wear on Monday?"

"I don't know."

"Wear one of your short, tight dresses. I know! Wear that blue one — the one with the scoop neck that falls open when you lean over."

I knew which dress she meant. It was Alex's favorite. All at once I was filled with guilt. I haven't done anything wrong,

I told myself. But I still felt shitty. Really shitty.

"And wear heels, not flats. Maybe you could even wear black nylons . . ."

"Stop it! Stop it!" I couldn't listen to her anymore. Grabbing my purse, I ran out the door.

73. WHY DO BIRDS MIGRATE?
The primary function of migration is to locate a more favorable environment, with regards to food, climate, and breeding place.

Alex came in quietly and unobtrusively, as always. I was on the floor, stretched out on my belly, listening to the first Food of the Gods album, reading my book about birds. I wasn't even aware he was home until he knelt beside me, gently tugged on my pony tail, and kissed me behind the ear. I turned over on my back and received a proper kiss on the lips. One thing about Alex, he was a great kisser. He applied just the right amount of pressure, and his lips were always nice and moist. Sometimes I tried to keep my eyes open when we kissed, but as soon as we made contact, they always fell shut.

"Home already? What time is it?"

"About 6:00. You have a good day?"

"I guess so," I nodded. I didn't really recall what I'd done that day. Nothing, really. Nothing that stuck out in my mind. "How about you? How was the store today?"

"Just your basic Saturday in the middle of the summer. Busy." He rose and wandered into the bathroom. I finished reading the question about migration, then got up. We met in the kitchen.

"What's for supper?"

"Salad for me. Had a big, fat lunch."

"What did you have?"

"Pizza."

"Luigi's?"

I nodded.

"Well, I didn't have time for lunch, and I'm starving." Alex opened the refrigerator and began to rummage through its contents. It always fascinated me the way he could eat a slice of cheese, a raw hot dog, then some fruit, and maybe even some cookies if we had any — all before dinner; and then hesitate when I offered him dessert afterwards. "I really shouldn't," he always said, patting his belly, which had gotten bigger since we moved in together.

That evening while I ate my salad he devoured a hunk of steak, a baked potato, and some rice left over from the night before.

"Kind of flowers did you send my mother," he asked in between gulps of milk.

"I bought her a plant. A fuchsia. It was really pretty."

"No flowers?" He stopped eating, and with his fork full of rice poised, he regarded me doubtfully.

"It's a flowering plant," I said.

"Oh."

"Really pretty."

"How did you sign the card."

"My name and then yours. Guess I lost my head."

"Guess so."

After I'd done the dishes and he'd settled down to watch *Jeopardy!*, I went into the bathroom. I saw he'd left the seat up. Why does he always do that, I wondered. I go out and buy a nice pink, fuzzy cover for the toilet, and he leaves the seat up. It occurred to me that I should speak to him about it instead of bitching to myself. But I doubted I ever would.

We went to bed early. I assumed we'd go right to sleep, because Alex seemed so tired. In fact, he'd yawned so many times that he finally felt he had to justify himself by saying, "I worked hard today. I'm beat." But once he hit the sheets, it was a different story. As he watched me pull off my Audubon tee shirt, he was suddenly wide awake.

"What, two nights in a row?" I giggled as he yanked me onto the bed with him. He didn't respond, just began to kiss one of my shoulders. It amused me to see him so transformed. When I'd told him earlier why birds migrate, he'd barely been able to keep his eyes open.

After he dropped off to sleep, I lay on my back, watching the cobweb, listening to the radio, and thinking about what Nick had said.

". . . A long-time resident of Boston plunged to his death today, from his ninth-story apartment . . ."

Was he right, was I unhappy with my life? Did I need some kind of change? Was it time to do something drastic? Something no one — including myself — would ever expect me to do?"

". . . a beautiful day tomorrow, with highs in the mid-eighties . . ."

But what would I do? Quit my job? Marry Bradley Crenshaw III? Go on a cruise? Plunge to my death?

". . . And now, join us for forty minutes of your favorite music . . ."

I sighed. Alex mumbled something in his sleep, then flopped over on his side. I stared at his back. It was nice and wide and still tan from when we went to the beach a couple of weekends ago. I reached out and touched his skin, which was very warm and soft. Absently, I stroked him, and he went,

THE DYNAMICS OF FLIGHT

"Mmmmm." I forced myself to yawn, anxious to join him in sleep. But as I reached over to shut off the radio, my hand stopped suddenly and I nearly fell out of bed.

"What the hell . . . this sounds like . . . but it can't be! They broke up!"

Squinting until my forehead ached, I stared at the radio. The song I was hearing sounded exactly like Food of the Gods!

"No way!" I cried, heedlessly waking Alex. Food of the Gods, a band that was formed when I was in my teens, had always been my favorite. I listened to them constantly, despite the fact that they'd stopped working together after the release of their last album, about five years ago. I kept listening, rigid with attention. That vocal . . . that *had* to be Daniel Parker. I shut my eyes and thought the bass line was characteristic of Food's Joey Porter. When it ended, I bit my lip and held one hand over my heart, waiting to hear those magical words that Food was back together. But the DJ went straight into another song, an inferior cover of Sam Cooke's "Wonderful World." Throwing back the sheets, I propelled myself out of bed and flew to the phone.

"Lark? What are you doing?" I heard Alex ask sleepily.

The line was busy. I hung up and dialed again. And again and again. But it was as if the phone was off the hook. I tried a dozen times to get through, but kept getting a busy signal. Alex appeared, still naked and slouched with sleep.

"Shit!"

"What's the matter?"

"Alex! I think I just heard a new song by Food of the Gods!"

"I thought you said they broke up."

"They did! But they must be back together again! Jesus! Wouldn't that be incredible?"

"Incredible," he agreed, stumbling back into the bedroom. I took a deep breath and tried the number again, but it was still busy.

"Call them tomorrow," Alex suggested. I tried one more time, then gave up.

"I will," I said as I got back into bed. "I'll call them tomorrow and ask what song they played at five after midnight."

"You do that," Alex sighed, and drifted off. I listened to him sleep and thought that if Food of the Gods had reunited, it would almost make me happy. And then I wondered why I wasn't already happy.

CHAPTER FOUR

The next day I called the radio station about a hundred zillion times, but each time the phone was either busy, or rang and rang and rang without anyone answering. In a flurry of obscenities, I finally gave up. Alex, listening to me curse and sending me sympathetic glances, seemed to be working up his courage to tell me something.

"WHAT," I said impatiently.

"You haven't forgotten, have you . . . that we're going to my parents' house today? To celebrate my mother's birthday?"

Hot anxiety shot through me. I'd forgotten, alright. Shit! My rage faded, replaced with despair.

Alex saw the look on my face and said, "Come on, Lark, she's not that bad."

"Yes she is!" I wailed. Unexpected tears filled my eyes. All of a sudden, I felt frighteningly as if I was about to go straight out of my mind. I felt dizzy with doubt and fear and despair. I felt like I was going to pass out, and wake up in a mental hospital wearing white. I could even feel myself trembling a little. I tried to stop, but couldn't.

"Lark," Alex said, alarmed, putting his hands on my

shoulders and studying me, "are you okay? You've been kind of on edge lately. Is something wrong?"

Kind of on edge? I almost laughed. Instead I just shook my head.

"Why don't you change, and then we'll go over. You'll feel better."

"Change?" I echoed. "What do you mean?"

"I mean . . . I mean . . . you're not wearing *that* to my parents' house . . . are you?"

We both surveyed my attire. That morning I'd thrown on a navy blue hooded marsupial sweatshirt, grey running shorts, and Nikes with pink laces that Alex had bought at his store for me. He was right. The outfit was hardly appropriate.

"What's wrong with what I'm wearing," I snarled.

"Well, I mean, I think you look fine. But I thought you might want to dress a little, uh, cooler. You'll probably be hot in a sweatshirt."

"No I won't. Your mother always keeps the air conditioner on too high. If I change into something nice, I'll be too cold," I said, childishly determined to wear the sweatshirt and shorts. Just the idea of dressing to please Alex's mother made me furious.

"Suit yourself," Alex punned weakly, seeing it was unwise to try to reason with me.

Mrs. Hawkins hated the fuchsia. She had it sitting on an old newspaper on the kitchen floor. I spotted the card in the garbage. It had gotten wet, and my name was smudged.

"I don't hold out much hope for it," she said. "I don't know much about exotic plants. Flowers are so much easier to take care of. You just put them in water, then throw them away when they die."

I looked at the fuchsia. It seemed to have wilted since I'd seen it last. Maybe it was starting to get to know Mrs. Hawkins, and was going to commit suicide.

"If it begins to lose leaves, I'll take it to our place," I offered, stooping to examine it. It needed to be watered. They say that anytime you re-pot a plant, or even transport it from one place to another, you should water it. I fingered a limp blossom. "I'm pretty good with plants, and maybe I could . . ."

"That's a good idea. In fact, why don't you just take it tonight, my dear? I don't think I'd be able to take care of it here."

Probably not, you old witch, I wanted to confirm. Alex shot me an anxious look, so I forced myself to swallow my exasperation, and joked that next year we'd send her a plastic plant.

"A visit from my boy — that's all I want!" Mrs. Hawkins purred, claiming Alex with a maddening glow and leading him into the living room to join Mr. Hawkins in a before-dinner beer. I was left with the fuchsia.

"In a few hours you'll leave here for good," I said enviously, "but I'll have to come back again and again and again and again." The prospect made me shudder.

Over dinner we discussed politics. Not exactly my strong point. Basically, I know who's President, but not much beyond that. I floated in and out of the conversation, mostly out. I was trying to decide where Alex got his looks — he didn't seem to resemble either of his parents. He was very cute, but not the kind of guy you would stop and gawk at on the street; he was the type that you fall in love with, and then realize he's the most appealing, adorable man in the universe. I smiled a little as I looked at his short, wiry, dark hair, his very very green eyes, and his nice straight teeth. He was laughing at something he'd said, and I smiled some more, because it was such a nice sound. I admired his chin, which was covered with dark stubbles, even though he'd shaved that morning. Then he shrugged his shoulders, and I shut my eyes and recalled what he looked like without a shirt. His chest was very wide and solid (he played football in high school) and had just the right amount of hair on it. A little sigh escaped me as I thought about the way I liked to rest my head on it after we've made love, and listen to his heart go THUMPthump THUMPthump.

"Did you hear me?"

"What?" I snapped guiltily to attention.

"My mother asked you a question, Lark," Alex said.

"Oh, I'm sorry. What was it?"

"I asked you, my dear, what you thought of the new tax laws."

Were there some new tax laws? I looked at my hands and thought about how Mrs. Hawkins never ever called me by name. She always called me "dear" or "my dear," and once she called me Ann, which was the name of Alex's last girlfriend. Alex let it slip once that his mother had adored Ann and had never really gotten over their breakup. Seems they were actually engaged to be married, and Mrs. Hawkins had picked out a gown to wear at the wedding and everything.

"Mom, you know Lark never thinks about things like that. I mean, you know how she is with money. She wouldn't even remember to deposit her paycheck if I didn't remind her every week."

Alex grinned his support. Thanks a lot, I thought as I saw Mrs. Hawkins frown.

"Well she *should* think about things like that! She should get her head out of the clouds AND THINK ABOUT THINGS

LIKE THAT!"

I was startled to hear myself accused of having my head in the clouds twice in two days. That's how people perceive me, I marveled. They think I'm residing on the outskirts of life. To my sudden alarm, a smile tugged at my lips. I couldn't help it, I was proud of that, "residing on the outskirts of life." I'd just that moment made it up. I missed Mrs. Hawkins' next remark, but I heard Mr. Hawkins tell her mildly to leave me alone. I noticed that he looked like Alex as he smiled at me.

"Oh yes, the new tax laws," I said. "A real problem. A *real* problem. No telling what's going to happen next."

Alex saw through the bullshit and pinched his lips together. Mrs. Hawkins was clearly vexed at my unexpected comprehension, and didn't answer me. Instead, she smoothed her dress over her large, middle-aged body, folded her hands in her lap, and crossed her legs at the ankle, all in one smooth, familiar motion. She reminded me of a gigantic hen sitting on a roost.

"Who would like coffee," she said.

"I would, I'll get it!" I offered, leaping out of my seat and scurrying down the hall that led to the kitchen. It was so nice to get away from her. My eyes fell upon the fuchsia, and I shook my head and whispered, "What a bitch!"

And then I heard Mrs. Hawkins say in an insufficiently hushed voice, "Alex, *really*. Something should be done about her."

I kept quiet and listened. I heard Alex's voice, but couldn't distinguish his words. He was probably sticking up for me. Then I heard Mr. Hawkins say, "Oh, Arlene, she's alright. Besides, Alex loves her. Isn't that what's most important?"

"I suppose so," Mrs. Hawkins agreed reluctantly, but the tone of her voice said, "How he can be happy with *her* is beyond me!" Hot tears welled in my eyes and I was sniffling as I began to prepare the coffee. Even though I didn't like Mrs. Hawkins, I could have, if she liked me.

"She's evil," I hissed to the fuchsia. "I'm so nice to her, and all she d-does is p-pick on m-me." A couple of tears rolled down my cheeks, and I allowed myself a self-pitying sob as I thought about how unfair it was.

Meanwhile, the coffee had begun to percolate. I reached into the cabinet to take out cups, but through my misty eyes I couldn't see very well. I didn't get a good grasp on one of them, and it fell to the floor with a crisp, delicate SMASH. For one nightmarish second I stared at the porcelain fragments at my feet. I heard Mrs. Hawkins say with poorly restrained fury, "Now what has she done? Has she broken one of my good cups?" and I was immobile with horror. I heard Alex say, "Lark? Are you alright?" and then I heard their chairs slide away from the table. Their footsteps entered the hall, and I heard them getting closer and closer. And in the split second before they rounded the corner and appeared in the doorway, I grabbed my purse and the fuchsia and flew out the door. I heard them calling my name. I heard Mrs. Hawkins say, "Oh will you just look at that mess!" and I heard Alex shout, "Lark! Where are you going?" but I didn't stop. I jumped into my car, and the fuchsia and I drove away just as fast as we could.

CHAPTER FIVE

For some reason, my first impulse was to go see my mother. We'd never had a very warm relationship — I was always much closer to my father. But when I left the Hawkins' house, all I wanted to do was sink into my mother's arms and have her comfort me. So I drove through a couple of towns until I reached her place, and pulled up to the front curb. To my dismay, I noticed an unfamiliar car in the driveway. That meant she had company. Hoping it was a girlfriend or a bridge partner, but knowing it wasn't, I stopped the car and got out.

The yard used to be beautiful. Bright flowers used to parade along the fence, and the shrubs were always immaculately trimmed, like neat beards on well-groomed men. But now, even though the grass is always dutifully mowed, my mother's yard lacks the vibrancy it used to have when my father took care of it. I think that's one reason why I don't like to visit her very much. Her yard is too sad.

Making my way up the walk, I heard two voices. One, of course, belonged to my mother. The other was masculine and unfamiliar. Normally I would have just walked in — I mean, after all, I used to live there. But something prompted me to knock instead, and await permission to enter.

In a moment my mother opened the door and regarded me with a puzzled, but welcoming smile. As usual, I was struck by how pretty she was. She's kept her figure trim, she applies makeup like a professional, and she's always dressed very stylishly. Dismally, I recalled that I was wearing that sloppy sweatshirt and shorts combination.

"Hi," I said.

"Lark! What a nice surprise! Come in." She stepped aside as I walked in. We hugged awkwardly.

Sitting on the couch was a large, red-faced, balding man, who rose heavily as I entered, wearing a big, friendly grin.

"Lark, this is James Chapin. We met through a mutual acquaintance."

That meant she was dating him. If it had been a woman, my mother would have said, "A friend introduced us." But since it was a man, they'd met through a "mutual acquaintance."

"I've heard a lot about you, Lark," James Chapin extended a damp, beefy hand, and I shook it. He asked if it was "hot enough" for me. I nodded absently and turned to my mother, waiting for her to begin some kind of conversation.

"So! What brings you by? Where's Alex? Would you like a glass of lemonade? James and I are having some." She indicated a couple of glasses sweating on cork coasters on the coffee table.

"Okay. I'll help you get it."

Our eyes met. Somehow she knew something was bothering me, and it wasn't just because I'd offered to help her. Much as I hate to admit it, we had a real mother-daughter thing happen at that moment. She nodded, told James we'd be right back, and escorted me into the kitchen.

"What's the matter?"

37

I didn't answer right away. She reached out, took my chin in one of her well-manicured hands, and made me look at her. I couldn't recall the last time I'd seen such concern in her eyes, and all at once I started to cry.

"Lark! What is it?"

"I h-hate my life!" I sobbed. She pulled me close, and I clung to her, getting her expensive blouse wet with tears.

"Oh for heaven's sake. Why."

I opened my mouth to explain, but no words came out. I didn't know why. I honestly didn't know what was so awful about my life. I had an okay job, a nice boyfriend, my health was good and my financial situation wasn't alarming. I felt myself flush with reluctant guilt the way I do when I complain that my hair is full of split ends, and someone else says they know someone who had leukemia and lost all their hair and then died. My life isn't so bad, I thought . . . but why am I so unhappy? Why do I hate to get up every morning? Why don't I ever smile for no reason, or sing to myself? I pulled away, shamed.

"Forget it," I mumbled.

"Are you getting your period?" my mother asked indulgently. Before she went through menopause she used to have fierce periods that turned her into a trembling, sobbing wreck. As a result, anytime I'm upset, she assumes it's because I'm getting my period.

"No, it isn't that. I'm just a little . . . I just came from Alex's parents' house."

"Oh." Her mouth tightened. She'd met Mrs. Hawkins, and didn't care for her any more than I did. "Where's Alex?"

"He's still there. I just took off. I couldn't take it anymore."

"Did it ever occur to you that if you would just marry Alex, she would treat you better?"

"That may be so, but I'm not going to marry him just to please her. Or you, either," I added sullenly. Sometimes I wondered if my mother loved Alex more than me, because it absolutely infuriated her that I wasn't married to him.

She bit her lip and released me. I felt bad being so nasty to her when she was trying to help, and I knew I should apologize, but for some reason, I didn't. I just looked at her and awaited her obligatory words of comfort.

"Well you poor baby," she said, and in a way it sounded unexpectedly sarcastic. But then she said brightly, "Have some lemonade. You'll feel better. You can visit with James and me. You'll like him — he's been to Europe. Okay?"

I nodded and wiped my eyes on the sleeve of my torn, navy blue hooded sweatshirt. She poured me a glass of lemonade. I sipped it and was disappointed. It didn't taste anything like the way it used to when my father made it. He used to put so much sugar in it that you could see about half an inch of it settled on the bottom. But my mother disapproved of sugar. I grimaced to let her know I hated it, then followed her into the living room. James Chapin rose again as we entered. My mother sat on the couch next to him while I dropped into the chair opposite.

"She usually dresses with more care," my mother announced suddenly, critically surveying my attire.

"No I don't," I said.

"She looks fine," James Chapin said heartily. He had a big belly, and his hands rested on it like a pregnant woman. His bald head shone in the sun coming through the window. My father always kept himself in shape, and his thick, grey hair always used to curl on humid days.

"Don't you think it's time you cut your hair, Daddy?"

"No way! I'm going to grow it long, like yours! Can I borrow one of your hair ribbons?"

"Okay! What color?"

"Hmm, let's see . . . how about red?"

"Red! Daddy! Red is a girl's color!"

"Ha! That's all you know! Some of the most beautiful male birds in the world are adorned with red feathers!"

"I'm sorry, what did you say?"

"James just asked you if you liked your job at the travel agency," my mother said.

I started to say, "Yes," out of habit. But instead, I took another sip of lemonade and gave his question careful consideration. Did I like my job? Did I like listening to Pat and Alice giggle and gossip all day? Did I like stinking of cigarette smoke after I got home at night? Did I like arranging trips for other people all day long?"

"No," I said, "I hate it. I'm going to quit."

My mother's eyebrows rose to the very top of her head.

"I must say, you're acting *very* strangely today," she observed. "What will you do after you quit?"

"I don't know." I didn't. The news had come as a surprise to me, too. "Might take some time off. Might travel."

"Can you afford to do that?"

"I don't know. But I'm going to, anyway!" Sudden hope surged through me. I leapt to my feet. "In fact, beginning right now, I'm going to start a new life. I am! Isn't that a great idea?"

"A great idea," James Chapin confirmed comfortably.

"It doesn't sound very practical to me," my mother stood, too, and with a sigh, James Chapin heaved himself to his feet.

40

"No, it doesn't, does it. But I don't care! That's what I'm going to do! Right now!"

And with a laugh so merry it startled even me, I soared out the door to begin my new life.

CHAPTER SIX

As the fuchsia and I drove along, I thought about that phrase, "a new life." I even said it out loud. "I'm starting a new life," I said. What did that mean, exactly? What would starting a new life entail? Would I change my name? Cut my hair real short? Read subversive literature? I wondered what other people did when they started new lives.

As I pondered this, I kept driving. I didn't intend to run away. I planned to go back to the apartment I shared with Alex, make some kind of plans, and then carry them out. But instead I just kept driving. Something inside was saying Go! and I was saying, Okay! So I drove and drove. What a good thing Alex and I had opted to take my car to his parents' house! Running out on him was one thing, but to use his car to do it would have been totally lacking in scruples. Besides, I liked my car better. It was a relatively new, metallic blue Grand Am with a dark blue interior. Rode like a dream, made any trip a pleasure.

And I suppose that's why it didn't seem like very much time had passed before I reached the Massachusetts Turnpike. I was startled to be confronted with the option of going to New York if I wanted to. And suddenly I wanted to. "Hello," I said to the man at the toll booth as I waited for him to give me my ticket. For the millionth time, I wondered how much a job like that paid. He didn't respond as he shoved the ticket into my hand. I said, "Thanks!" very brightly, a little hurt by his failure to greet me, and pulled away, heading west.

When I reached the Sturbridge exit I got off and followed 84 for a while. Jersey barriers were lined up to the left of the fast lane, and they unnerved me. I was afraid I would drive into one, simply because it was there. In the meantime, it was getting kind of dark. So I turned up the volume of the radio and sang along with a Steely Dan tune. For the first time, it occurred to me that even though I knew all the lyrics, I wasn't sure if the song was about breasts or fish.

And then I was welcomed to the state of Connecticut. I was counseled, as well, regarding the use of radar detectors. Seems it's perfectly legal to own them, but you're not allowed to detect radar with them.

I kept driving for a couple of hours or so, and before I knew it, I was entering Troutskill, New York. I had no idea how I'd gotten there. I mean, if someone had asked me that morning, "Say, do you know the way to Troutskill?" I would have said No. But suddenly there I was, and what's more, I was glad. Unexpected memories flooded all my pores, as all at once I remembered I'd been there years and years ago with my father. We'd just taken off one weekend without bothering to tell my mother.

"See that bird there, Sweetie?"
"That black one?"
"Yep. But look again. See? It isn't really black . . . it's purple and blue and black. Isn't that something?"
"Ooh, he's pretty! What kind of bird is he, Daddy?"

43

"He's a grackle."
"He sure is shiny."
"He sure is. His feathers look just like your hair — like iridescent satin."

While I drove, my hand was on the radio dial, searching for stations playing the song I'd heard the night before, the one that sounded just like Food of the Gods. In a funny way, I wasn't thinking about how I was running away, leaving my old life behind. I was thinking about everything else. I was listening very intently to the radio. I was trying to discern what I could of the Troutskillian countryside in the darkness. And I was thinking about my father.

I stopped at a money machine, used my card to gain access, pressed a couple of buttons, and effortlessly had enough money for food, gas, and whatever else I might need.

"One helluva country," I said to the fuchsia as I slid back into my seat.

I stopped at the next gas station, filled my tank, then went off in search of some place to eat. On the way, I noticed that the town had really grown up since I'd been there. The King Kone where my father and I had gotten ice cream was gone, replaced by a video rental shop. That kind of made me sad. With a sigh, I wondered what time it was. I wasn't wearing my watch because it had abruptly ceased functioning earlier in the week, which had really annoyed me at the time — I've always been one of those people who absolutely *have* to know what time it is — but which now seemed insignificant. I figured it was probably about midnight. I wasn't at all tired. It was still hours before I'd feel like I'd be able to sleep, because of my insomnia. Once Alex asked me why I couldn't just relax and let myself fall to sleep. I told him it's because I don't like to sleep.

I always feel like I'm wasting my time, or that something really exciting will happen and I'll miss it. I told him that my fear is that someday I'll look back on my life and realize I never did anything exciting.

Bright orange letters that said **D NER** caught my eye. I pulled up to the curb, stopped the car, and went inside. It was one of those really cheesy places, with red and white grimy tiles on the walls, and a well-worn, grey tiled floor. I was silently urged by the management to seat myself, which I did, choosing a booth near the window, through which I could keep an eye on my car. The seat of the booth was covered with red plastic that was supposed to pass for fabric, and there was a long tear down the center that was sloppily secured with grey duct tape. I tried to avoid it, but wound up getting some of the adhesive on the back of my thigh. While I tried to rub it off inconspicuously, I saw there was an ancient juke box on the table that probably hadn't operated properly since the mid 70s. I gave up trying to get the adhesive off my thigh, and flipped through the metal tipped pages. "I Got You, Babe," "Saturday in the Park," "The Cover of the *Rolling Stone*." I finished reading the selections, but the waitress was still nowhere to be seen, so I started reading them again, this time more carefully, and saw "Young at Heart," an old Sinatra tune that was one of my father's favorites. I remembered that he used to sing it to me and say, "Frank is right. It's important to have lots of dreams. That way if one of them doesn't come true, you have plenty of back-ups. Don't you think that's a good idea?" I used to nod earnestly and say, "Yes, Daddy, that's a real good idea." And sitting there in that diner in the middle of the night, I realized I didn't have *any* dreams.

"Ready to order?" I heard a weary voice say suddenly. I looked up and saw an alarmingly thin woman regarding me listlessly. She'd brought me a glass of water, and I saw that most of the ice had melted. The menu she handed me was typed and had mistakes that were corrected by hand. For some reason, nothing appealed to me.

"Just coffee, please," I said.

She nodded, reclaimed the menu, and disappeared. I sank my chin into my hands and sighed deeply. What the hell was I doing there? Why wasn't I at home, listening to Alex sleep?

The waitress spilled my coffee as she served it, but didn't apologize. I had to use my only napkin to mop it up. She wore a name tag that said "Melanie."

"Thank you, Melanie."

"Welcome," she said without moving her lips, slid the check under my saucer, and disappeared.

Maybe she's having a bad day, I thought. Maybe she's getting her period. I sipped my coffee. It was bad. It was too dark, and I added too much sugar, which gave it the consistency of syrup. I took another sip and counted the dead flies scattered along the windowsill. Seven and a partial. As I fished in my purse for change, I came upon the button that Nick gave me, the one that said, "Life sucks, then you die." Instead of leaving a tip, I left the button.

I stopped at the first motel I saw, a cruddy little run-down place unjustifiably crowned The King Arthur Motel. The parking lot was deserted. I hoped the place hadn't been condemned or something. It seemed like a viable possibility. But there was a light on in the office, so I went inside. Seated behind a desk was a man I presumed to be the manager. On his desk was a sign that said, "Curtis Henderson, Manager," so I took a chance and greeted him by name. He was watching a tiny black and white television, and I recognized an old *M*A*S*H* rerun, the one where Hawkeye and Hot Lips share a kiss while their shelter is being bombed.

"How you doing," he said.

I told him I was fine, and that I'd like a room for the night. He nodded and handed me a key that had a piece of masking tape stuck to it. The tape had a five written on it in red Magic Marker. He asked if the day had been "hot enough" for me, and I said it had been.

"Check out time is 11:00."

"Okay, I'll be out by then."

"Well . . . if you're not, that's okay." He became affable all of a sudden, contemplating me with a sleepy but kind smile. I guessed I must have looked pretty forlorn, standing there holding my fuchsia and a key with a red five on it. "My wife and I usually have coffee at 11:00. You're more than welcome to join us for a cup."

"Thanks, I'd like that," I said gratefully. I was pleased. I've always believed that for every unfriendly person, there's a friendly person, and that it's good luck to meet them on the same day. It was something my father had told me once.

"But why is it lucky to meet them on the same day?" I wanted to know. Outside the wind was making the windows in the kitchen rattle, and somewhere in the yard we kept hearing a woody THUMP, like maybe the door of the shed had blown open. But inside my father's study it was safe and warm and dimly lit. He was sitting at his desk, and I was standing very close to him. Our eyes were at the same level, and we were both smiling and listening to Mother prepare dinner.

"Well, Kitten, because then you can relate one to the other. In that way, you can understand that for everyone who's unfriendly, there's someone who's nice."

"But why do there have to be unfriendly people at all?" I

persisted. It wasn't so much that I was interested in his answer; I just loved listening to him talk. I loved the sound of his sweet voice, and I loved having his gentle brown eyes looking into mine.

"Well, if you never met an unfriendly person, you wouldn't appreciate someone who was being friendly, now, would you."

"I guess not."

"See, you can't have the good without the bad. Know what I mean? You can't have warm weather all year long, because then you wouldn't appreciate the sun. You'd take it for granted, and then you would never have those moments when you take a great big breath of fresh air, listen to the birds sing, and think, What a beautiful day! See what I mean?"

"Uh huh."

"I don't know about you," he said, drawing me close and kissing my head, "but I wouldn't give up those moments for anything in the world!"

"Me neither, Daddy."

"We'll see you at 11:00, then," Curtis Henderson said.

"What? Oh. Oh yeah, okay."

"Have a good night."

I assured him that I would, and left. I was feeling pretty good as I made my way up the narrow walk to the door that had a five on it. I let myself in, turned on the light, and surveyed the room. It was exactly what I'd expected — motelishly barren except for a bed, a nightstand, a desk, and a TV. I didn't turn on the TV because watching *M*A*S*H* would have made me sad. Used to be my father's favorite show. I noticed that the TV had a radio built into it, so I turned that on instead. Going into the bathroom, I tore the paper off a glass on the sink, filled it with water, and brought it out to the fuchsia.

"Suppertime," I said, pouring it into the dirt.

Then I pulled off my Nikes, my sweatshirt, and my shorts, and climbed aboard the bed, which was hard and made up with very stiff, unfriendly sheets. I left the radio on and pretended that Alex and I had just made love and that he was asleep in my arms. I was lonely for him, but I'd walked out on him, so I had no one to blame but myself. Forcing myself to think about other things, I lay awake for nearly two hours before I finally fell asleep.

Around dawn I awoke without knowing why. Then I became aware that the breath-stoppingly beautiful voice of Daniel Parker was filling my cruddy little motel room.

"The new Food tune!" I cried, leaping from the bed and scampering over to press my ear against the radio's tiny, tinny speaker. But as soon as the song ended, another song took its place, a real lame one full of clichés about how painful love is. Then the DJ came on and announced that it was going to be another scorcher.

"So call in sick or quit or do whatever you have to do to be able to spend the day at the beach!"

"Just tell me the name of the song, goddammit," I muttered. But instead, an affray of commercials assaulted the airwaves. I listened long enough to get the name of the station, and then I tried to call, but the phone in the room had been disconnected.

"Figures," I grumbled, crawling back into bed. But I couldn't sleep. So I lay there, waiting for the sun to come up, listening to the radio and wondering what I was going to do now that I had a new life.

CHAPTER SEVEN

Pretty soon the room grew uncomfortably warm, and the heat brought out some kind of weird smell from the bedspread. I got up and took a long shower, then climbed back into my sweatshirt, shorts, and Nikes. Meanwhile I was still listening to the radio, and everyone was saying that it was going to be a magnificent day. I looked out the window and decided to take a walk and let my hair dry in the sun. For no apparent reason I've always been under the impression that if I let it dry in the sun, it will get lighter. Once I even squirted lemon juice on it to aid the process. But of course it never got any lighter. Unless you want to count those grey hairs. Seems the older I get, the less my hair resembles the iridescent plumage of a grackle.

More changes had taken place in Troutskill than I'd noticed the night before. It was a lot busier. Some of the streets that had been quiet and friendly, now had cars parked on both sides. A Dunkin' Donuts had taken the place of the town's little library, where my father and I had gone in to look up the Loggerhead Shrike in Peterson's classic *Field Guide to the Birds*. Still, Troutskill had maintained its small town charm, and as I walked, I greeted people who were on their way to work. Most of them smiled, and probably about half of them asked if it was "hot enough" for me. I kept saying Yes, but it's a question I've never understood. The first time I heard it, I asked my father about it. He said it was just something people say to be friendly, which seemed strange to me, because I could think of all kinds of more effective ways to be friendly. But anyway, now every time I hear it, I think of my father.

I was burning up in my dark sweatshirt, so I went back to the motel. At the door of the office I heard voices, and called out, "Hello!" The sun was very bright, and as I went inside, everything was draped in a cooling darkness. I couldn't see anything but outlines.

"Here she is," one of the outlines said. "This is my wife, Loreen."

I nodded at the outline that hadn't spoken, and it said, "Nice to meet you."

"Nice to meet you," I parroted.

"Sleep okay?" inquired the first outline, who was by now showing signs of detail.

"Yes, thanks. My fuchsia is still in the room. I'll go get it."

"No need to hurry. Have some coffee."

"Thanks. My name is Lark, by the way."

"I know. That's the name you signed last night."

"Oh yeah."

"So. Been on the road long?"

"About seventeen hours, I guess," I said. Curtis and Loreen both began to say, "Wow," when they heard that number seventeen, but when it was followed by "hours," not "months" or "weeks," their voices faded off, and they didn't know how to respond. I felt badly, like I'd disappointed them.

"Vacation?" Loreen asked after a moment.

Vacation? I almost laughed.

"Kind of," I said.

"Where you headed."

"Not sure. I'm just kind of driving."

"By yourself?" As I nodded, Loreen placed a hand over her heart and exclaimed, "Heavens! I'd be scared to death! Aren't you scared?"

"No." Being scared was something that hadn't even occurred to me. I had too much other stuff to worry about. Like where was I going, what was I going to do when I got there, and if I would be happy once I arrived.

"Sounds a little funny to me," Curtis observed frankly.

"I suppose so," I said.

"Not in any trouble, are you?"

"Not that I know of."

"Not fleeing from the police or anything like that? Because if you're fleeing from the police . . ."

"Curtis, stop that," Loreen interrupted peevishly. "She's not *fleeing from the police*. Honest to God, you watch too much TV."

"Still sounds funny to me," Curtis insisted. Loreen clicked her tongue and shook her head. By now my eyes had grown accustomed to the darkness, and I saw that she was a sturdy woman, short and solid, with dark grey hair and a very kind face. Curtis, too, looked kind. I mean, even though he was accusing me of being a fugitive from the law, he was doing it in a nice way, like he wasn't going to turn me in. I felt a warm affection for them both seep into me.

"I ran away from home," I finally explained.

"I knew it!" Curtis cried triumphantly. "I said to myself, A girl like that is running from something."

"Oh *Curtis*," Loreen said.

"Trouble with your family?"

"My boyfriend. I just walked out. I guess that means we broke up." I felt sad hearing myself say that, and a gloomy resignation settled over me, replacing the excitement I'd felt earlier. I wondered what Alex was thinking. Was he worried about me? Was he mad? I'd never known him to be mad before, not even at unreasonable customers, not even at this woman who came into his store one day while I was there, demanding a 20% discount on her purchases because her daughter worked at the main office in Rhode Island; however, she had no proof of this, so Alex told her he couldn't give her the discount. Well the woman exploded! She said she was going to call the main office and speak to the president of the company and see that Alex was fired that very day. Alex, who knew his job was under no threat, apologized anyway, and told her he'd give her the discount. Furthermore, he told her, if she ever had any problem with any item ever purchased from the store, she was to ask for him personally, and he'd take care of her. She melted at that, said he was a very sweet young man, and went away satisfied. Impressed, I asked Alex how he could be so nice to her when she'd been such a bitch, and he said, "First rule of retail — the customer is always right." I asked him how come I was never right when *I* was the customer, and he said it was because I insist on shopping in low class department stores.

"Don't you worry, honey," Loreen's gentle voice broke into my thoughts. Handing me a cup of coffee, she motioned that I should sit on a chair. It had a blue plastic seat, and I knew it would make my thighs wet with sweat. I sat anyway. "If things are meant to be, then they *will* be."

I pondered her cryptic message as I sipped my coffee. It was weak and sweet, just the way I like it. And as soon as it hit my stomach, a slow rumble like prolonged thunder erupted and filled the office. Curtis and Loreen stared.

"Guess I'm a little hungry," I said sheepishly. "I haven't eaten since I left home."

"But that was seventeen hours ago! You poor thing!" Loreen went over and crouched before a small refrigerator in the corner that I hadn't noticed before. A second later, she withdrew a couple of voluptuous peaches.

"Would you like a peach?"

I nodded eagerly. I've never liked peaches, because I hate that fuzziness in my mouth. I'm more a nectarine fan. But the peaches she was holding looked good. Damn good. She handed one to me. It was chilled and soft. I took a huge bite, and juice dribbled down my chin. Loreen and Curtis smiled broadly.

"Not much of a breakfast," Curtis pointed out. "We got anything else?"

"Got another peach. Have it," Loreen urged.

"No, no, this is fine. Honestly. I'm trying to lose a little weight anyway."

"If you lose weight, there won't be nothing left of you!" they said, aghast. Then Curtis' eyes narrowed.

"That why you left your boyfriend? He told you to lose some weight?"

"No, he'd never say that," I said at once. Alex always told me I was beautiful. "He would never say that," I repeated.

"Well if you don't mind my saying so, I think you still love this guy. Why'd you leave him, anyway."

"Curtis! That's none of your business."

"Just wondering."

"I don't know," I answered. I bit of silence ensued, uninterrupted except for my chewing. I hated eating in front of them, especially when it was so quiet. I tried to eat without making any noise, but the peach was so juicy that I kept having to kind of slurp it. I wished I was eating a nice solemn banana. When I finished, I lobbed the pit into a metal wastebasket next to the desk, where it ricocheted with a loud CLAK that startled us all, even though we were expecting it, and fell inside. I rose. Curtis and Loreen smiled.

"Guess I'll get my fuchsia and be on my way," I said.

"Good luck, Lark," Curtis reached out and shook my hand.

"Thank you. Thank you for the coffee and the peach."

"Take the other one, too," Loreen begged, and after a second, I accepted it, juggling it from one hand to the other while we swapped Goodbyes. Then I collected my fuchsia, waved at the two of them standing in the doorway of the office, and drove off.

"Nice couple," I told the fuchsia as I bit into my second peach. "My father would have liked them."

It was too bad he wasn't there with me.

CHAPTER EIGHT

Still following 84, I left New York and entered Pennsylvania. All the while I was fussing with the radio and thinking about dumb stuff. If I was stranded on a desert island with only five record albums (and a stereo system) which would I want them to be — Food, early Queen, probably not any classical, although you never know, sometimes you're really in the mood for it; if I could select five men I'd most like to be stranded on the island with me — Daniel Parker . . . Gregory Peck . . . David Letterman . . . maybe Tom Cruise, who technically was too young for me, but hell, we were stranded together on a desert island, so I mean, how choosy could he afford to be. I also gave some thought to what Curtis and Loreen had said. I supposed it was true, I supposed I still loved Alex. It would be impossible not to.

I met him on a blind date. The funny thing was, he wasn't *my* blind date. My then-current boyfriend, Martin, had fixed him up with my best friend, Linda. Alex and Linda seemed to get along okay, although Linda was obviously more interested than Alex. As for me, I wasn't happy with my relationship with Martin. He was very domineering, and always criticized my appearance. In fact, that afternoon we'd had a gigantic fight about my hair. He said, "You're not going to wear those retarded barrettes, are you?" I was hurt, because I've always been sensitive about my hair. I've always noticed that whenever I go out with other women, I like their hair better than mine. Mine is too thick, and it's difficult to manage or keep under control. It doesn't hold any kind of style, so I usually keep it out of my eyes with these colorful barrettes. I guess it's true, I guess they're kind of childish. But they remind me of some barrettes my father bought me once when I was a little girl.

Anyway, Martin hated them. He kept insisting I get my hair cut and professionally styled. But I didn't want short hair. My father used to tell me that all the best looking women have long hair, and he made me promise never to cut it. "If you really loved me," Martin said that afternoon, "you'd wear it the way I want you to wear it." "If *you* really loved *me*," I said, "you wouldn't care how I wear it." "You should spend more time on your appearance," Martin said, and I said, "What? And be vain like you?" because he always sported a very chic hair style and wore expensive clothes. "A little vanity wouldn't kill you," he said. We glared at each other. I left the barrettes in, and the ride to the restaurant was pretty quiet.

When we arrived, Alex and Linda were already there — he'd picked her up right on time. "He's such a gentleman," Linda hissed while he and Martin greeted one another. "He opened my car door for me and everything."

We ordered, and then Linda and I excused ourselves and went to the ladies room so that we could talk some more.

"Don't you just love Alex's green eyes?" she said.

I said I hadn't noticed his eyes, I was too pissed at Martin. I told her what he'd said, and she said he was an asshole. We returned to our table, and right away I noticed that Alex had the most beautiful green eyes I'd ever seen.

Over dinner it became apparent that Alex and I had a lot in common. We liked all of the same movies, we read a lot of the same books, and we even discovered that our jobs were comparable, because we both had to deal directly with the public. We laughed and swapped rude customer stories, and generally ignored our dates. I felt sorry for Linda, but I couldn't help it, Alex was so cute and so funny. Martin and Linda, who never liked each other to begin with, began to bicker, and as soon as the bill was paid, we all left. Alex drove Linda home. Martin and I went back to his place and resumed our fight. I told him I never wanted to see him again, and I even said I didn't want to be reincarnated and see him in my next life, either. I mean, I was *really* mad. He said that was fine, and refused to drive me home. So I had to walk six miles.

The very next day Alex called me and asked me out. I moved in with him three weeks later.

That was two years ago. I wondered what had gone wrong.

I took Route 380 to Scranton, then got on 81 headed south. By now I was getting hungry again, so I exited at the first sign that indicated food, and pulled up to a family restaurant.

A pleasant woman with obviously dyed hair greeted me and asked me if I wanted Smoking or No Smoking. I said, "No Smoking," and she escorted me to a table next to the window. She made a comment about the weather as she handed me my menu, then departed politely while I made my choice. By now I was so hungry that there wasn't anything I didn't want to order about three of. I took a long time making my decision — I had to send the waitress away twice — before I finally opted for blueberry pancakes. Incredibly fattening blueberry pancakes. While I waited for them, I sipped my water and glanced around at the other patrons. A couple nearby bickered about real estate. I could tell the woman hated the guy, because she was glaring at him and grimacing each time he called her "baby." His hand rested arrogantly on her knee, and she kept glancing down at it with a displeased frown. Totally oblivious to her petulance, he kept talking, on and on and on. I felt too sorry for her to look at her anymore, so I allowed my gaze to wander over to another couple who were evidently deeply in love. They looked as if they couldn't wait to finish their meal and go some place where they could be alone. Watching them, I felt about a hundred years old, thinking, I used to be like that once, which made me really sad. I looked away, hoping to see something that would take my mind off my so far wasted life. Something outside the window caught my eye, and I saw a huge crow so black it looked like a silhouette, picking at a piece of bread. My father told me once that crows can live up to thirteen years. Crows in captivity can live even longer — up to twenty years. I recalled hearing that some crows in laboratories have been taught to "speak," and count to the number four.

While this corsuvian trivia was going through my head, a sleek grey pigeon appeared, hoping to join the crow in his meal. The newcomer had those friendly bland eyes, and he approached the bread with just the right amount of respectful hesitation. His neck shimmered like oil on pavement. Talk about your long life spans — pigeons can live up to thirty-five years. The crow granted the pigeon a couple of shy but determined pecks, then took some sedate steps toward the bread. The pigeon retreated, stood at a slight distance and waited patiently. The crow stabbed viciously at the roll, then backed off. The pigeon then moved in for another bite, then the crow, then the pigeon, and so on.

My pancakes arrived. I buttered them and squirted maple

syrup on them. At the same time, the real estate guy and the woman were rising to leave. She was wearing an expression that was a combination of boredom and rage. He was shaking his head with a patronizing "Women! Can't live with 'em, can't live without 'em!" kind of smile that bugged the hell out of me. I frowned and eased my fork through all four layers of pancakes. The other couple was still deeply engrossed in one another. They not only held hands, I saw, but their feet were touching, too, as if to maintain as much physical contact as possible. It reminded me of the way Alex and I used to sit sometimes, and I had to turn away again. While the first bite of pancakes melted in my mouth, I looked back out the window. Immediately, my eyes flew open.

A huge guy with long, black hair and a snaggly, bushy black beard was flinging open the door of my relatively new, blue Grand Am!

"That's my car!" I hollered, leaping up from my table and scrambling to the door. His eyes met mine for the briefest of seconds, and I saw that they were tiny and cruel. Instinctively I knew I'd never be able to stop him, so I said the first thing that came to my mind, which was, "My fuchsia is in the front seat!"

I heard him roar with the most sinister laugh I'd ever heard, and then my fuchsia came flying out the window. As he sped away in a billow of gravel and dirt, it landed with an explosive SPLOT at my feet. I slammed my eyes shut to keep out the dust, and when I opened them a second later, they were brimming with tears. I knelt down next to the fuchsia and frantically began to replace some of the soil that had spilled, scratching my hands on the pavement as I scooped up the dark, rich dirt. By now the tears were streaming down my cheeks, and my nose was running. Each time I reached up to wipe my eyes, I left a line of wet, snotty dirt.

Immediately I was surrounded by a gaggle of sympathetic waitresses.

"Oh, your *car*! What a *shame*!"

"Did you get a good look at the guy who took it?"

"You can use our phone to call the police."

I was too stunned to be grateful. I just squatted in the parking lot, blindly attempting to re-pot my fuchsia.

"Here, we'll take care of that." I felt someone take my arm and draw me to my feet. Someone else picked up my fuchsia. I was led solemnly back inside.

"Why don't you go into the ladies room and clean up," suggested one of the waitresses. "We'll call the police for you. By the time you get out, they'll be here, and they'll help you find your car."

I nodded like a child accepting instruction, and walked dejectedly into the ladies room. I knew what the chances of recovering my car were. Zilcho.

When I caught sight of my reflection in the mirror, I burst into fresh tears. I looked horrible! I was wearing day old clothes, no makeup, and my uncombed hair hung in dark snarly strands. My hair had come out of the confines of those childish barrettes. Not only that, my cheeks were smeared with potting soil. With a shaky sigh, I splashed water on my face, unsnapped my barrettes, attempted to pull a brush through my hair, and noticed that I'd gotten pancake syrup on the pocket of my sweatshirt.

The door opened and one of the waitresses said in a kind voice, "The police are on their way. You feeling a little better?"

"Oh yeah," I snapped, "I feel just fucking *great*. Some asshole just took my car!"

She could have been offended, but she wasn't. She said, "I know," and came over and put her arm around my shoulder in a comforting half embrace. I looked at the two of us in the mirror and wanted to cry some more, but there really wasn't time. She released me, then held the door open and stepped aside.

"Where's my fuchsia? Is it alright?"

"It's fine. Ellie took care of it."

As we joined the others, I saw her send them a look to warn them I might become hysterical. That was how I felt, like throwing myself on the floor and kicking and wailing. Like a single, motherly unit, they fluttered around me, patted my shoulders, led me to a seat, and offered me coffee and more pancakes. I shook my head. Someone handed me my fuchsia, and I accepted it wordlessly. I sat unhappily with it on my lap, thinking that except for my purse and my clothes, it was all I had. While I waited for the police to arrive, I inspected the damage. Several of the delicate blossoms had fallen off, and it had lost a lot of dirt.

"It's been traumatized," I said. "It's going to need some water."

The waitresses scattered, occupied with my plaintive request. I was suddenly left alone. Absently, I tapped the pot in my lap. Tap. Tap. Tap tap. Tap tap tap. Taptaptaptap. Quicker and quicker; and then my foot began to tap, too. Fury was replacing my shock and sadness, and I felt like I was going to explode. What kind of person would take another person's car, for Chrissakes? How was I supposed to run away from home if I didn't have any transportation? Shit! I thought, is this what Nick meant about experiencing reality? By the time the squad car swung into the parking lot, I was roasting in rage.

CHAPTER NINE

The first cop who climbed out was tall and muscular, with a stern, angular face, very short hair, and mirrored sunglasses. He had that tough, unshaven look that a lot of women really go for, but it didn't do anything for me. The other cop stepped out a couple of seconds later, and I saw that he was young and looked inexperienced. He wore aviator style glasses and had a thin, timid moustache. I met them at the door, still clutching my fuchsia.

"Some son of a bitch stole my car!" I wailed, and to my horror, burst into a round of girlish, inconsolable tears. The first cop yanked me into a masculine clutch and said, "Okay, little lady." From behind me, I heard a couple of the waitresses go, "Oooh!" The other cop glanced back at them, then regarded me with sympathy that didn't quite strike me as being genuine. He was very thin. He didn't look like he should be a cop, he looked more like he should be selling art supplies in some little shop downtown.

"Gimme all the details," growled the cop who was holding me tighter and tighter. He smelled overwhelmingly of musk, and I was suddenly reminded of Alex's fragrance. Alex never used cologne or anything, he just had a smell that was characteristic of him. It was a combination of his shampoo, the detergent I used on his clothes, his deodorant, and the natural scent of his skin. Once he had to go to an overnight convention for drugstore managers, and I slept with one of his unwashed undershirts.

Wriggling out of the cop's squeeze, I said there wasn't much I could tell him. I could see dual reflections of myself in his sunglasses, and I looked frantic and sloppy.

"I can describe my car," I offered.

"Didn't you see the guy?"

"Kind of. He had tiny, mean eyes, and a dark, bushy beard. He was the most horrible man in the world," I said.

"Not much to go on," observed the smaller cop wryly. "What're we gonna do, post signs that say, 'Wanted: The most horrible man in the world?'"

"Pipe down," the bigger cop admonished him smartly. The little cop glared at me, resentful that I'd made him look bad in front of all the waitresses. The big cop went on, "You'll have to come down to the station and file a report."

"Shit," I said.

The waitresses circled me and patted me comfortingly on the shoulders, but their eyes were glued to the big cop. I seemed to be the only woman there who wasn't willing to give up a year of life to spend just one night with him. They told me I didn't have to pay for my pancakes since my car had been stolen.

"Very nice of you ladies," said the big cop, introducing himself as Officer MacKenzie, and putting his hand on my back as he opened the door of the car for me. I hesitated. He'd opened the front door.

"Shouldn't I sit in the back?" I said.

"Nah. Whadya want, everyone to think we've taken you

prisoner?"

There were appreciative giggles from the waitresses. To hear them you'd think they were watching the Letterman show, for crying out loud. I didn't laugh as I got in.

"I'll sit up front, too," the little cop said, promptly climbing in after me. And that struck me as *really* strange. But I didn't question it. Maybe that was how they did things in this town.

Officer MacKenzie circled around, back to the driver's side, and let himself in. Sandwiched between the two of them, I sighed and turned around to watch the restaurant disappear from sight as we drove off, then faced forward again. I'd been to a police station once before in my life, one afternoon when my father and I were on one of our trips. We'd broken down by the side of the road, and a gang of teenagers had pulled over on the pretense of helping us, but they'd taken all my father's money instead. When they drove off, my father and I hitched a ride to a gas station, and called the police from there. A very nice officer, Officer Walker, showed up a couple of minutes later and took us back to the station. He hadn't been able to get our money back, but he'd been so nice that my father shook his hand vigorously for a long time, and patted him on the back, and invited him over to dinner the next time he was in our neighborhood.

"So what's your name, little lady," Officer MacKenzie asked.

"Lark DePaolo," I said.

"Glad to have you aboard, Lark. By the way, this here's Officer Willard."

With a jerk of his head he indicated the thin cop, who nodded sullenly, still smarting from having been chewed out in front of all the waitresses back at the restaurant.

"So where do you live."

"Massachusetts."

"Yeah? What're you doing here. Visiting?"

"No. I'm just on a trip."

"Uh huh. What kind of trip."

"Just a trip. I just felt like, you know, taking off."

"Yeah? Why's that."

Why is everyone so suspicious? I wondered, recalling that Curtis had accused me of fleeing from the police. Did MacKenzie suspect that I was somehow involved in the theft of my own car? I felt like I was in the middle of a bad movie made for TV.

"Look, don't you want to know about my car?" I demanded.

"Sure. Tell me about your car. Whadja do, leave the keys in the ignition?"

"I did *not* leave the keys in the ignition!" I said, "The window was down, and that guy just got in my car and hot wired it."

"The most horrible man in the world," Officer Willard reminded us. I turned and sent him a stare so snotty that he looked out the window. Officer MacKenzie, meanwhile, continued with the investigation. He was driving with his left hand on the wheel, and his right hand resting innocently on the seat, about an inch from my thigh.

"What was this guy wearing."

"Jeans. And he had on a blue shirt . . . I think . . . I don't know! I'm sorry . . . I was just so surprised that I c-couldn't s-see him very w-well." I started to cry again, I was having a real cruddy day. Officer MacKenzie's hand rose, landed on my thigh, and patted it gently.

"Don't you worry, little lady. We'll find your car."

"How," asked Officer Willard. His eyes were on MacKenzie's hand. I was very conscious of my short shorts.

"What will you do if we don't recover your vehicle? Do you have friends you can stay with?"

"No."

"Want to get a room at a motel?" he asked, then turned red as a Cardinal. "I mean," he said stiffly, trying to ignore Officer MacKenzie's appreciative guffaw, "can we take you to a motel to get a room. For yourself, I mean . . . Christ! Shut *up*, MacKenzie!"

"That's not how you make time with a lady, Willard!" Officer MacKenzie crowed, tapping my thigh twice — once for "la" and once for "dy."

"*You're* the one who's trying to 'make time' with her," Officer Willard pointed out, injured and humiliated. I couldn't help feeling sorry for him, but I didn't say anything. I just sat there between them, feeling very shy and holding my fuchsia with both hands.

An unexpected sharp turn sent me careening into Officer MacKenzie's lap. He said, "Hold on to something, darlin'!" and it sounded very sexual to me. We pulled into the station and came to a stop. I straightened up with as much dignity as I could, and noted with some satisfaction that his neatly pressed navy blue slacks were partially covered with potting soil. Officer Willard got out, then leaned over and reached in to take the fuchsia. I handed it to him, then scrambled out, aware that Officer MacKenzie was watching me from behind.

"Here you go," Officer Willard said, returning my fuchsia. Standing close to him, I saw he was a little older than I'd thought before. Grey threaded through his sandy brown hair, and his eyes looked weary, like someone who'd been through a lot. I wondered if maybe he was a Vietnam vet, and decided to be as nice to him as I could. So I gave him a big smile as I thanked him.

67

"Cut the chit chat, Willard," Officer MacKenzie snapped as he joined us. "We got a damsel in distress here."

Taking my arm, he escorted me up two steps and through a door that had the words **Police Station** stenciled on it. We were met by a gust of incredibly hot, stale air. There was an ancient fan by the window whose vigorous WIRRRR sounded unjustifiably effective. Three tired officers, all 50 or older, looked up as we entered, then swapped significant glances that I didn't understand.

"What's the problem," asked one of the officers whose badge, I saw, said "Officer Holloway."

"This little lady lost her car."

"I didn't lose it, some bloody thug *took* it!" I said, pissed beyond belief. Officer MacKenzie smiled approvingly at my outburst.

"Know something, little lady? You got spit. I like that."

"I'd like to spit on *you*," I said. "Stop calling me 'little lady.'"

"He calls all the girls 'little lady,'" explained one of the cops with a placid smile.

"Meant no offense," Officer MacKenzie affirmed humbly.

I scowled. My hands closed tight around my fuchsia. Officer Willard, who I hadn't noticed was gone until he came back, said, "Lark, we've got an APB on your car. In the meantime, why don't we get started on this report. You want a cold drink or something?"

"Please," I nodded gratefully.

"Must be pretty warm in that sweatshirt," observed Officer MacKenzie. "Why don't you just take it off?"

The cops all laughed. I turned to Officer Willard, but at the same time he reached out to hand me a can of Coke. Gracelessly, I bumped my chin on it.

"Thanks. I owe you," I said. I didn't mean it to sound like I was going to have sex with him, but it evidently sounded that way to the other officers, because they hooted. They were like teenage boys who couldn't think of anything but women's naked bodies. I thought back to Officer Walker, the officer who'd helped my father and me, and recalled that he'd bent down, I think I was about six, and tugged on one of my pig tails. *What a pretty little girl you are. Your father must be very proud of you.*

With a mild frown, Officer Willard seated himself at a typewriter. I dropped into the chair across from him.

"Name?" he said, even though I'd already told him.

"Lark DePaolo."

"Place of residence."

I didn't answer right away. Since I didn't live with Alex anymore, I really didn't have a home.

"Place of residence?" he repeated, looking up from his typewriter.

"Don't have one," I admitted.

He studied me. Then he sighed and said, "Look. I need to put an address here."

After a minute, I gave him my mother's address, and that seemed to satisfy him. Then he asked me to tell him exactly what had happened. So I said I'd been eating blueberry pancakes, and I'd looked out the window just in time to see a big bearded guy climb into my car. Officer Willard typed and nodded. He asked me some more questions, and then yanked the form out of the typewriter.

"Now we just have a few pictures for you to look at," he said. From a shelf behind him, he pulled out a book so enormous it reminded me of the photo albums my girlfriends showed me when they got back from their honeymoons. With a deep sigh, I opened it and began to flip through the pages.

CHAPTER TEN

"None of these guys look like the one who took my car," I finally reported several hours later. Officer Willard looked up from some paperwork he was occupied with, and heaved a sigh.

"I was afraid of that," he said. "That reduces our chances of recovering your vehicle."

"Well, so what do we do now?"

"How about going out to dinner with me."

I was astonished — not that Officer Willard had asked me out, but that he'd asked me out in front of the others.

"How about taking her to a motel, too," Officer MacKenzie sneered. He was watching me like a hawk with all the others, awaiting my response. I didn't have that much desire to go to dinner with Officer Willard, but I sort of wanted to spite all the other cops. Besides, it wasn't like I had all kinds of other options open to me.

"Okay. Thanks."

Officer Willard couldn't keep the triumph off his face.

"Nothing fancy, you understand," he added hastily and sheepishly. "This job doesn't pay that well."

"I'm not dressed for anything fancy anyway," I reminded

him. As I shut the album and pushed it aside, then rose and stretched, his eyes took in once again my sweatshirt, shorts, and Nikes.

"You look fine to me," he said. But there was nothing lecherous about his voice or expression.

"Where shall we go?" I asked brightly, slinging my purse over my shoulder and picking up my fuchsia.

"Willard, for Chrissakes! You can't take her out to dinner!" Officer MacKenzie exploded. "She's come to us for help! She's been victimized! It's unethical to take advantage of her vulnerability."

"It's unethical for me to take her out for something to eat, but it's okay for you to grope her in the squad car?" Officer Willard demanded. His thin, timid moustache quivered indignantly. The other cops chuckled. Officer MacKenzie swore under his breath and called Officer Willard a "peckerhead."

"Let's go," I said, ignoring them all. My plant and I walked out through the door, down the two stairs, and out to the parking lot. After a moment, Officer Willard joined us.

"Be careful, Lark," I heard Officer MacKenzie warn.

"What a jerk," I marveled.

"You can say that again," Officer Willard agreed.

We went to a little French café full of round tables covered with red and white gingham table cloths. The waitresses all wore white ribbons in their hair, and were dressed in very short, very perky black dresses with white bows, like French maids in an old Hollywood movie. *Trés chic.*

A croissant appealed to me more than anything else in the whole world. I relayed this to Officer Willard as soon as we entered. He nodded, and a hostess whose name badge said

"Suzette" seated us.

"Think I'll have the Special," announced Officer Willard. I looked it up in my menu and saw that it was *poulet*. Chicken. As I nodded my approval, I couldn't keep from telling him I'd taken six years of French in school.

"That so?" he murmured, still scanning his menu, as if he wasn't quite convinced he'd made the best choice.

"*Absolutement*," I said. "Think I'll have a croissant. *Frommage*."

"That all? Not much of a dinner."

"Well, I'm trying to lose a few pounds."

It would have been sweet if he'd expressed surprise, but he didn't. He just nodded and said he'd always been the type of person who had to *gain* weight.

"So how long have you been a cop," I said, changing the subject.

"Not long . . . and not for much longer. I'm more interested in architecture. Been taking courses at night."

"Really? Good for you. When will you be able to quit this job and start doing what you really want to do?"

"Not sure yet. I . . . what's wrong?" he demanded, seeing that my face had suddenly been crossed by lines of intense attentiveness. I was sitting rigid in my chair, staring at something above his head. "What is it? What's wrong?" he repeated, turning around to see if he could determine what I was looking at.

"Shhh!" I hissed. "*Listen!*"

We strained our ears. Sure enough, from somewhere in the kitchen I could hear . . . was it? Yes! The new Food of the Gods song!

"Holy shit! Excuse me, will you?" I jumped from my seat. My napkin fluttered to the floor and our glasses of water rattled.

Officer Willard rose too, awkward and baffled. "No, no, you *stay*!" I cried, inadvertently giving him the command I used to give my dog Elliot when I wanted him to sit on the porch while I went out to get the mail. Officer Willard was startled into obedience. I hurried off to see if I could get into the kitchen.

One of the waitresses intercepted me with a helpful but puzzled smile.

"Help you?" she asked.

"I need to get into the kitchen," I said urgently, trying to get past her. But she blocked my way.

"Why?"

Meanwhile I was still listening to the song. It was very upbeat. From what I could hear of the lyrics, it was about a guy who thought he was "*at the bitter end/without a single friend*," until he discovered that his love of music made his life worthwhile.

"God, it's great, isn't it?" I shook my head.

"What's great?"

"That song! That song coming from the radio in the kitchen! *Shhh!*"

Just as she quieted down, a voice from behind said, "Excusez moi, I am zee manager. Is zere a problem?"

I turned and saw a tiny man dressed in black slacks and a crisp, white shirt. His accent sounded forced, but I had to give him credit for trying.

"*Pas de tout,*" I said, "Not at all. I just wanted to . . . wait, wait, wait . . . it's over! No one say anything!"

They complied, without knowing why. The three of us turned our ear to the kitchen and heard, *"Ah, what a nice song that is. That's the new one from . . ."*

"Lark! What the hell is going on?" Officer Willard suddenly boomed from behind. The manager and the waitress glanced at him, startled to have been joined by a man of the law, and one who was kind of pissed, at that.

"That song! I think it's Food of the Gods! I mean, I'm sure it is! It has to be!"

"What are you talking about?"

They were all staring. In fact, every patron in the café was gaping at my behavior which could only have been termed *gauche*.

"I think I need to use the ladies room," I stammered, feeling as if my face would burst into flames.

There was a moment of indecision. I could tell the manager and the waitress were debating whether to give me directions, or just kick me out for being crazy, possibly dangerous. Since I was with Officer Willard, they probably assumed I'd committed a crime and was under his custody.

"Over there, first door on your right," the waitress finally told me when Officer Willard apparently had no objection.

"Thanks. Be right back," I said, and hurried off to the first door on my right.

Once inside, I leaned heavily against the sink, released a giant breath, put my hand over my heart, and thought back on my recent scene. A week ago I'd never even have left my seat, let alone try to storm the kitchen. Feeling a little dizzy with exuberance, I turned around, and, expecting to see my usual pleasant face, was amazed at my reflection. I was a real mess, with my tangly hair and stained sweatshirt and no makeup, as if I'd just crawled out of bed. I'd achieved a real *savoir dormir* look. But what startled me was my eyes. They glittered in a way that I'd never seen before. I couldn't help admiring them. For the first time in my life, I felt like I looked wild and mysterious. My thick, uncombed hair reminded me of a gypsy. And I actually looked thinner. All of this shot through my mind in about ten seconds. Then I leaned toward the mirror and said to my reflection, "Wonder if the album is out yet?"

When I returned to our table, Officer Willard rose and said, "Don't bother to sit. We're not staying."

"Oh." I was startled. And hungry. "How come?"

"Tell me something. What're you doing in Pennsylvania by yourself?"

He had a nasty, suspicious look on his face. For some reason, it pleased me. No one knew me, no one knew how dull my life really was, they all thought I was running from trouble.

"I told you, I'm taking a trip," I said.

"Plan to stay here?"

"I don't know. It's funny, but for once, I'm not thinking in terms of the future, or what I'm going to do. I'm just acting on instinct, like an animal. And you know what? I haven't felt this good in a long time . . . not since my father and I used to take trips together. We never made plans or looked at maps. We'd just jump in the car and go. Used to drive my mother crazy. We'd disappear for days without telling her. That's probably why she wanted to divorce him."

"Oh."   Officer Willard was surprised by my lengthy response. As we stood there, next to our table with the red and white gingham table cloth, my mind went back to the day I first heard about it.

"A divorce? What are you talking about? I love you! I can't live without you!"

"Don't lie to me! You live without me whenever you take Lark and go away!"

"But I'm always thinking of you! I mean, you're on my mind, even though I'm not with you." My father's voice sounded frantic. On my perch at the top of the stairs I could

*picture his face, white with sudden, unexpected alarm, his
fingers fluttering in that nervous way of his.*

*"That's not good enough," I heard my mother say coldly.
"You're just not responsible enough to be married and raise a
child. You're a bad influence on her. I know you think it's fun
to pick her up at school and take her on a trip for a week, but
she's falling behind. Have you seen her report cards? I spoke
to Mrs. Ingram, her history teacher. Lark won't be going on to
the tenth grade. She's failed."*

*"But she's learning other things, important things," my
father said plaintively. I was surprised to learn that I wasn't
going to be promoted to the next grade, and almost missed my
father's next words, "She's learning about nature. She doesn't
care about some war that went on a hundred years before she
was born!"*

*"You know why you're not capable of raising a child?
Because you're a child yourself! And I refuse to watch over
both of you! It's too much! Don't you know how scared I get,
when I come home and find you've disappeared again?"*

"Scared? Of what?"

*"Scared that something will happen to you . . . that you'll
get into an accident . . . that I'll never see you again."*

"I'm sorry, what did you say?" I jumped at the sound of
Officer Willard's voice.

"I said, Does your husband know where you are?"

"I'm not married," I said. "I had a boyfriend, but I left
him."

"Have a nice evening," bade the manager, suddenly
appearing and motioning us to the door as if to say, "Get ze
hell out of my establishment."

I couldn't meet his eyes as I hurried out. Officer Willard

was right behind me. I climbed back into the car, and gently took the fuchsia onto my lap. By now it was dark out. It was time I started thinking about where I was going to spend the night. I mentioned this to Officer Willard as he got in beside me. And to my horror he said in a voice so quiet and steady it was sinister, "You're staying with me tonight."

"What? I am? I mean, I don't . . ."

"Shut up."

As he started up the car, I stared at his face. Suddenly he looked cold and mean. I was so completely startled that I couldn't even move. I was even more surprised when my door flew open, and someone dragged me out by the arm.

"You bastard! Don't you ever get tired of this shit?"

I struggled out of the grasp. The voice sounded familiar, and I recognized it just before I whirled around and saw Officer MacKenzie standing there. I was so glad to see him I could have wept. He reached into the car, retrieved my fuchsia, handed it to me, and told Officer Willard to take off.

With a dark scowl, Officer Willard drove angrily away. I stood there holding my fuchsia with my eyes and mouth wide open.

"I can't believe it! How did you know . . .?"

"Just a hunch," Officer MacKenzie said, and he sounded arrogantly confident, like Jack Webb on the *Dragnet* series. "When it comes to women, he's dangerous."

"I thought *you* were the dangerous one," I said clumsily, then slammed my mouth shut.

"It's the quiet ones you gotta watch out for. Ain't no one ever told you that?"

"But how did you know we'd be here?"

"Willard always comes here."

"Oh."

"He give you the line about wanting to be an architect? Taking courses at night, and that crap?"

My gasp was his answer. He shook his head.

"Bastard," he said. My heart was pounding. I hugged my fuchsia to my chest. Imagine, being rescued from Officer Willard by Officer MacKenzie!

"Come on, let's go back to the station. We can make some calls. Get you a place for the night," he offered kindly. I nodded breathlessly, and noticed that he *was* kind of handsome. Not adorable like Alex, but rugged and authoritative. As I climbed into his car, I glanced back and saw that everyone in the café had come outside to watch the action. I realized I still hadn't had anything to eat.

"When all of this is over, I'm sure going to be thin," I said.

CHAPTER ELEVEN

We drove back to the station. When we went inside I saw that only one cop remained. Officer Holloway. He was sitting at his desk doing some paperwork. Officer MacKenzie greeted him with an affirmative nod.

"Willard never gets tired of that charade, huh?" Officer Holloway chuckled. "The great architect!" Officer MacKenzie delivered a brisk smirk. He was still wearing his sunglasses with the mirror lenses, despite the fact that the sun had set hours ago.

"Yep. Rescued this little lady for the second time today."

I sighed. I didn't feel I was in any position to tell him to stop calling me "little lady." It was true, he *had* rescued me twice.

"Do you have a phone book? I should probably call some . . . where are you going?"

To my alarm, Officer Holloway was shutting off the light above his desk. As I watched, he recapped his pen and headed toward the door.

"Home," he answered with a sly grin. I looked at Officer MacKenzie, but he was hunched over some reports, apparently hard at work.

It was an ugly situation for me. Naturally, my first impression was that Officer MacKenzie had set this up so that he could be alone with me. But he looked so innocent that I felt guilty for thinking such a terrible, ungrateful thing. I mean, for crying out loud, he was an officer of the law. But wasn't Willard an officer of the law, too?

"Have a nice night, Officer Holloway," I said doubtfully.

"You too, *little lady*!" he said with a laugh that made me uneasy, and disappeared. I listened to him go down the steps, then I heard him open his car door, slam it, start up the engine, and drive off. I jammed my suddenly damp hands into the pocket of my sweatshirt and said, "How about that phone book?"

Officer MacKenzie looked up from his work.

"Phone book? What do you need a phone book for?"

"To find a motel. I need a place to stay tonight."

He smiled so slowly that I had scads of time to grow very apprehensive.

"Honey, you don't need to stay at no motel. Plenty of room here."

"Here?"

"All our cells are empty. They got beds in them."

"That's awfully tempting, but I don't think . . ."

"'Course we got a bed in back for when officers need to take a load off. Nice bed. Built for comfort. Big. Huge, in fact." I watched him stand, and saw that he was nearly a foot taller than me. From the look on his face I realized he wasn't talking about the bed at all.

"I really don't think . . ."

"Look. What do you think I'm gonna do, rape you? Goddammit! I won't touch you!" He was indignant and impatient. His feelings were hurt. Perhaps I'd misjudged, perhaps I'd be perfectly safe with him. I picked up my fuchsia and tried to think. Meanwhile he was coming closer and closer.

"I won't touch you," he repeated, "unless you *want* me to."

Reaching out, he caressed my cheek gently. I started to object, but he put a finger on my lips to shush me. "Let's get this show on the road," he murmured. Then, to my horror, he leaned forward, intent on kissing me.

RRRRRRRING! went the phone.

"Shit!" said Officer MacKenzie, taking a step back toward his desk. I spun around and flew out the door and heard it go SLAM! behind me. Officer MacKenzie called me back, but I knew he couldn't come after me. I scrambled down the two steps and started to run. It was awkward, because I was still holding the fuchsia, and with each step a little more soil spilled out. I wasn't used to running, and my breath was rushing in and out so fast it hurt. My chest began to throb. Finally I had to stop. But I kept walking quickly, until I reached a convenience store that was still open.

"Thank goodness," I whispered. "I can go in and use the phone to . . . hey!" Stopping dead in my tracks, my mouth dropped open. Parked in the lot was my blue Grand Am!

I was *sure* it was mine. Glancing quickly around for the bearded guy and not seeing him, I scurried up to it and looked inside. Sure enough, I saw my sunglasses on the dashboard and the "Jesus Loves You" pamphlet some woman had given me a few days ago. The door was unlocked. "He should know better than to leave it open in this neighborhood," I muttered, putting down my fuchsia, retrieving my keys from my purse, and climbing inside.

As I was reaching down to grab my fuchsia, I heard someone holler, "Hey! Get the hell outa my car!"

It was him! My hands broke into a violent tremble, but still

I managed to jam the key into the ignition. My Grand Am started right up. I saw the bearded guy galloping toward me. I heard the THUD THUD THUD of his footsteps as he came closer. I wondered what my funeral would be like. I wondered when my body would be discovered, and what condition it would be in. Would it be suitable for viewing, or would it have begun to decay? What would mourners say about me? What kind of things would they tell Alex, to make him feel better? Would Pat and Alice be able to attend, even though it would mean shutting down the travel agency for a couple of hours? And what about Alex's mother? Would she feel she was in some way responsible for my death? Probably not. She'd probably be glad, the bitch.

"Hang on!" I told the fuchsia as I backed out with a roar. I could hear the guy still shouting, but fortunately I couldn't distinguish the words. I shifted into drive with a bone-jarring THUMP, and sped off. In my rear view mirror, I saw him running after me, shaking his fist. He grew smaller and smaller, then disappeared as I rounded a corner.

"Thank God, thank God, thank God," I said. I was shaking so badly that I could barely drive. My first instinct was to stop at the first motel I came to, but I was afraid the bearded guy would come after me, so I drove through several towns. I turned on the radio, hoping to hear the new Food tune, jumped onto the Interstate, and continued to drive until it started to grow light out.

Finally I exited, and stopped at a sleazy-looking establishment that offered special hourly rates. I tried not to think about that as I made my way to the office. By now I was so tired I could hardly walk.

A man with a thin ring of red hair and a bad complexion napped in a chair at his desk. He looked mean and ugly, like

Sinclair Lewis in a bad mood. I cleared my throat timidly. His eyes opened slowly, then he sat up straight.

"Help you? Need a room, honey?"

"Yeah."

"For how long?"

"One night."

"That long, huh?" He chuckled.

"I'm alone," I said.

"A shame. Guess you won't be needing the Tarzan and Jane room."

"Guess not."

"Number seven, then. Nice room."

"Fine."

As he handed me a key, he told me how much I owed him. I paid him and left the office.

"Number seven," I said to my fuchsia as I fetched it from the car and carried it to the room with me.

I made sure the door was locked and all the shades were down before I pulled off my sweatshirt and shorts. I even took off my panties, and washed them in the sink with a small, hard, fragrant bar of soap that didn't produce many suds. Hoping they would be dry by the time I put them on again, I hung them in the shower. I watered the fuchsia again and promised I'd get it some more soil as soon as I could. Finally, I threw my weary body onto the bed. It was a softer mattress than I like. Probably the springs were worn out from lots of action.

I'd no sooner shut my eyes than I heard a knock on the door, and a voice hissed, "Lark? Lark? It's me, Stan."

"Stan?" I sat up in alarm, clutching the sheets tightly around me. My eyes were open so wide they immediately started to ache.

"The manager."

"What do you want?"

"Just wanted to make sure you didn't, you know, *need* anything."

"No, I'm fine," I said as firmly as I could. I wondered if he had a key to number seven. Probably. I hoped he wouldn't use it. I was too exhausted to fight him off. I could see his silhouette through the curtain, and I could tell he was trying to peer in. He had his hands cupped around his eyes, and he was hunting for the split in the curtains, hoping to get a look at me. But of course all the lights were out, so he couldn't see anything. I was so tense I felt like I was going to throw up. But finally he left.

"Jesus, what a night," I sighed, and tried to relax. But the sun had begun to shine through the curtains before I finally slept.

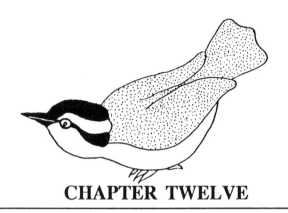

CHAPTER TWELVE

A harsh KNOCK KNOCK KNOCK woke me from a dead sleep. I sat up, blinked, and didn't have any idea where I was.

"Who is it?" I asked, awaiting recollection of the night before. I was in a real low-life motel. But what was I doing there? In a couple of seconds, it came back to me. Oh yeah, I'd narrowly escaped getting raped twice, and I'd recovered my car all by myself, and then I'd somehow managed to avoid getting raped again. I was exhausted.

"Rise and shine, sugar, it's check-out time."

I recognized Stan's voice. He was miffed that I'd rejected him. With a deep, lethargic sigh, I dragged myself out of bed.

"Stan, could I have a few extra minutes, please?" I begged him from my side of the door. "I overslept. I'm really sorry."

There was some silence. I think I dropped off again as I waited with my head resting against the door, because I jumped a little when he said sternly, "Okay. Stop by the office with your key in one hour."

"Thank you, Stan." An enormous yawn undermined my gratitude. I went into the bathroom to take a shower, and discovered a cockroach in the tub. I was so alarmed that I did nothing but gasp and stare for several seconds. Then I turned the water on high, and shut it right off, hoping he'd get swept away by the water and spin down the drain. But he was too big. He didn't even bother to struggle as the water carried him, and when he landed at the drain, he just climbed off and headed for higher ground. Probably went through that every morning. While I watched him, I wondered if I was going to be sick.

"How about getting out of there and letting me shower," I said. "You've got the whole rest of the world outside to play in. Come on now, let's go."

But he ignored me. He just kept crawling on his horrible little legs toward the back of the tub. His long antennae were wet, and they kind of stuck to his back. Gathering up all my courage, I yanked on the toilet paper roll until I'd accumulated six or seven feet of it, then I scooped him up, and flushed him down the toilet.

"Can't say I didn't ask you to leave nicely," I said.

My shower was brief. I kept looking all around for more cockroaches, and my skin tingled as if they were crawling on me. I couldn't wait to get out of the room. So I dressed quickly, fluffed up my hair with my fingers, grabbed my fuchsia, and went to return my key.

I was debating whether or not to tell Stan about the cockroach, but when I entered the office, I saw a woman there instead. I assumed she was Stan's wife, and I couldn't help pitying her.

"How did you sleep," she said, unmindful of the sympathetic smile I was giving her.

"Fine, thanks. You?"

Instead of answering, she took the key and said, "Have a good day."

"Okay, thanks. You, too."

I stepped back outside, into the bright sunlight. It was

going to be another hot day. The fuchsia and I climbed into the car.

"Well! Wonder what's in store for us today!" I chirped brightly. The fuchsia regarded me sorrowfully. "Hey, you need more soil. Let's see . . . where could I get some . . . ?"

Meanwhile I'd started up the car and was backing out. My mind was racing. I needed to stop at a bank. I needed to get gas. I needed to get soil. And I needed to keep driving.

"We're going to North Carolina, you know," I said abruptly. Even I was surprised by my announcement. But I suddenly knew that I wasn't going home until I'd seen a Painted Bunting.

"Have you ever seen one, Daddy?"

"No, Sweetie, I haven't. But someday we're going to drive down to North Carolina, and we're going to see one together. Of all of my goals, that's top of the list. I want to see a Painted Bunting before I die."

But he never did. And I think I'd had it in my mind all along to drive south. I think I felt like I'd be accomplishing my father's goal for him, since he'd been unable to himself. I was on a mission.

Sobered considerably by this thought, I drove an hour before I stopped at a diner for breakfast.

"Don't let anyone take the car," I told the fuchsia sternly as I got out. I debated whether or not to leave it unlocked with the windows open. That was how it had gotten stolen the day before. The practical side of me advised locking it, even though it would be insufferably hot inside within a couple of minutes. My fuchsia wouldn't be able to take the heat. "Oh hell," I said out loud, "what are the chances of having my car stolen two days in a row?" I noticed I'd been doing that a lot lately, talking out loud to myself like that. I left the windows open.

Inside the diner it was cool. Huge fans on the ceiling created a breeze so strong that napkins fluttered on tables and ladies' skirts blew lazily. I took a deep breath while I waited to be seated. Outside the window I saw a couple of tiny House Sparrows and an Eastern Bluebird. The bluebird caught my attention because for a little while they were kind of rare in New England. The Audubon Society had made an effort to increase the population by erecting bird houses built specifically for their slight build. In fact, back at Alex's apartment somewhere I still had a diagram I'd cut out of *The Boston Globe*. I'd always planned to build one myself, but had never gotten around to it.

"I said, Are you coming, Miss?"

I was startled out of my ornithological observations by a hostess who was staring at me expectantly. I smiled sheepishly and nodded. Turning, she led me to a table along the back wall. She was slender on top, but across her bottom her white uniform was so tight that I could see the outline of her panties. In fact, I could even see that they had some kind of design on them . . . blue or purple zig zags. How about that. My eyes shot guiltily back to her face as she turned around and indicated that I was to sit. I thanked her, accepted my menu, and sat.

I couldn't see my car, but that was okay. I was next to a mirror, so I could covertly watch the other patrons. In fact, I got so involved with studying an elderly man with thin, white hair and long ear lobes, that when the waitress arrived, I had to confess that I wasn't ready to order yet. Instead of saying, "Okay, I'll give you a few more minutes, then," the way waitresses are supposed to, she frowned and stormed off. "How rude," I murmured, riddled with guilt just the same. For some reason, it made me think of the time Alex and I were buying window shades in a department store. We went up to the register, and the girl there, she was probably about sixteen or seventeen years old and chewing gum with an aggravating SNAP SNAP, said with a rapid Boston accent, "Cash, check, or *chahge?*" I waited for Alex to say "Charge" but instead he just smiled his gentle smile and said, "Whatever happened to 'hello?'" The girl was so surprised that she didn't know what to say. For some reason, that incident has always stayed in my mind.

In the meantime, I'd noticed that I'd been given a lunch menu. It must have been after noon. Too late for breakfast. It took me a moment to accept and adjust to that, and eventually I opted for the fresh garden salad. If I keep this up, I thought, I'll probably lose about twenty pounds. I was so excited at the prospect that I wasn't even tempted by the dessert menu. I relayed my modest order to the waitress the next time she appeared, and off she went to fetch it. I looked back in the mirror at the old, long-lobed man. He was sitting alone, engrossed in a book. I couldn't read the title, but it had a clock on the cover. He must have sensed I was watching him, because he looked up suddenly. I gave him a broad smile. He had a long, drawn face, and very gentle eyes behind small round glasses. He reminded me of a picture of Hermann Hesse my father kept tucked in his copy of *Beneath The Wheel*. After a moment, he returned my smile, in that shy, guileless way old men have.

Just then my salad arrived and my attention was diverted. I was dismayed to see that most of the dressing was plopped on top of the onions (which I'd specifically requested none of) so that when I removed the onions, I also removed most of the dressing. The rest of the salad was a bed of lettuce and a tomato, quartered and attractively displayed at each corner of the square plate. Instinctively, I knew it wouldn't fill me up, so I asked the waitress to please bring me a large order of fries, too. So what if I only lose ten pounds, I thought.

The salad was gone before the fries came. In the interim, I looked back in the mirror, back at the old man. He'd resumed reading. I couldn't imagine what his book was about, so I cleared my throat, and when he looked up, I asked him. Normally I wouldn't have. If I'd been sitting in Luigi's and I'd seen someone reading, it wouldn't even occur to me to speak to him. But things are different when you're on the road. Once Alex and I went to Lake Champlain for a week, and while we were boating we met a big Italian guy named Tony, and didn't think twice about inviting him back to our cabin for spaghetti.

The old man regarded me, startled but pleased, and held up his book, which I saw had the word *Horology* across the cover.

"Huh," I said, to indicate that the title hadn't satisfactorily answered my question.

"The science of measuring time," he explained willingly. "I repair clocks at a clock museum."

"A clock museum?" Until that moment, I'd had absolutely no idea that such a thing existed.

"Oh yes. If you've never visited it, you really should. It's fascinating. We have early lantern clocks and pendulum clocks dating back to the 17th century, and we even have an original Peter Henlein."

"No kidding, an original Peter Henlein?"

"You've heard of him?"

"Um, I'm not sure. Who was he again?"

"Peter Henlein created the first portable timepiece. A milestone in horology."

"How do you like that," I said. My fries arrived, and I

gave the waitress a smile of thanks. Meanwhile, the time guy kept talking.

"I'm going over there after lunch. Why don't you come with me? I'll show you around. You really should see it, you'll be amazed," he promised. I dipped a fry in catsup, bit it, and tried to think. Son of a bitch. A clock museum.

"Okay, I'll follow you over. Thanks," I said. I really had nothing else to do.

"My name is Professor Allen. May I sit with you?"

"Sure." This guy was set on fire by clocks. What else was he going to tell me? I gestured for him to bring over what was left of his meal, and introduced myself.

"Very nice to meet you, Lark," he said, setting his plate and silverware on the book and carrying that in one hand, and his coffee in the other hand. He spilled some on the way over, but didn't seem to care at all. I could tell he was really excited about telling me stuff about clocks. Once we were settled, he plunged into an oration on the history of horology. I tried to listen, just in case I ever found myself on *Jeopardy!* and one of the categories was "Time," but I was having trouble concentrating on what he was saying. I kept thinking how funny it was that some things you never ever think about are what occupy someone else's mind almost all the time. It's like the way I know about birds, and most people can't tell the difference between a Starling and a Brewer's Blackbird.

". . . which actually didn't strike the hour — in fact, they lacked both hands and dial — but simply contained an alarm to alert monks that it was time for prayer," Professor Allen was saying happily. "Those were your turret clocks."

"Uh huh," I said. I finished my last three fries (having saved the biggest and the greasiest for last) while he explained the mechanism of modern timepieces. I let most of what he said flow past me, nodding thoughtfully every now and then. I used to date an engineer, and it fascinated me the way he could talk for hours about something he *had* to know didn't interest me; I mean, he *had* to have guessed from my expression that I wasn't listening. Anyway, Professor Allen had a nice voice with a peaceful quality that nearly put me to sleep. He had a subtle British accent, like maybe he'd lived in England for a couple of years, and I enjoyed hearing it so much that it was easy to pretend I wasn't missing a word.

When the checks came, he paid for them both.

"Really?" I was touched and surprised.

"Been a long time since I bought a lady lunch," he said. He rose first, and with his hand, indicated that I was to precede him out. I heard him thank the hostess in his warm voice. I started to feel really good inside, the way I do when I see someone being nice just for the sake of being nice.

"I'll drive slow enough so that you can follow me," he assured me. "The museum isn't far from here."

"Okay. See you."

With a pair of waves, we separated. I got into my car. Where's my dirt? the fuchsia seemed to demand as I took a seat next to it.

"Oh quit whining for once in your life," I said. "You'll get your dirt. Soon as we see this clock museum. Imagine that? A clock museum!"

I let Professor Allen pull out ahead of me. He stuck his hand out the window and gestured that I was to follow him, which I thought was kind of silly, since we'd already agreed I would. But I waved back, and swung out of the parking lot after him.

CHAPTER THIRTEEN

Professor Allen was one of those old guys who drove an average of seven miles below the posted speed limit. Normally if I'd been behind him I would have honked my horn and shouted, "C'*mon*!" even if I wasn't really in any particular hurry. I don't know why, but I hate knowing that I could be going faster than I am. But now he was a friend of mine, and for once, I didn't mind driving slowly. So what I figured was probably a ten-minute drive took us twice that. And pretty soon I soon saw the sign for the clock museum, and we pulled into the driveway. He climbed out of his car and approached me leisurely. I got out too, and met him half way. Together, we headed up the steps.

"Here we are," he announced cheerily. "Originally I worked only a couple of days a week, but the museum's grown so much that I now I work full-time. On Mondays, all I do is wind clocks. Takes a full eight hours."

"Holy smokes," I said, "eight hours."

We made our way inside. There was an elderly woman with hair white as a Snowy Egret sitting at a desk. She looked too old to still have a job. Still, I admired her for allowing her hair to go so white. So many women insist on dyeing their hair brown, even when they're in their eighties. She greeted us with a gracious smile.

"Hello there, Professor Allen! Who's this?"

"This is my date," he grinned, "Lark DePaolo. I met her over lunch. She's interested in our clocks."

"Well isn't that nice. I hope you enjoy your visit with us."

I assured her I would, and, still smiling, she dismissed us. She looked very busy. I wondered what a secretary at a clock museum would have to do.

I followed Professor Allen into the main room, and was astonished to see that clocks covered every inch of space — on tables, walls, and the floor.

"Holy shit!" I exclaimed tactlessly.

"Let me take you on a tour," he offered with a gentlemanly inclination of his head. And slowly, we began to walk. He pointed out different timepieces, explained what year each had been made and who had made it, and chatted about time in general.

"Thoreau once said, 'Time measures naught but itself.' An intriguing concept, don't you think?"

I nodded absently. My father was a big fan of Thoreau, and I'd heard that before. I was fascinated by all the clocks, and my neck was craning this way and that in an attempt to see them all.

"How long did you say you'd worked here?" I was pretty sure he'd already told me, but I hadn't really been listening.

"Almost thirty years."

"Almost thirty years," I repeated, thinking that was how old I was. He'd begun working there the same time I was born. The coincidence struck me as being somehow significant. I fixed my gaze on him. "I'm about to turn thirty," I said.

"How about that!" He, too, seemed to think it was

meaningful. I waited for him to say, "Gee, isn't that too bad," or something. But instead he shook his head and said, "I envy you."

"Envy me?"

"Sure. Your life is just beginning."

"Just beginning?" I couldn't believe it. What was he, senile?

"You've probably got about fifty years ahead of you," he said comfortably. He took a couple of steps, eager to resume his lecture, but I didn't move, so he came back and studied my expression. "You're not happy about that?

"No! I mean . . . Jesus, fifty years! I'll be *eighty*! I'll be ugly and old and crippled . . ."

"I'm eighty," he pointed out, not quite amused, but not offended, either.

"Oops. I mean, well it's different for a man." I'd told Alex that, too. "What I mean is, all my best years are over. I'm not growing up anymore, I'm just growing old. I'll never be as pretty as I used to be. I'll never be as thin. My skin will never be as smooth. My hair is getting grey!" By now I was in a panic. Verbalizing my fears like that made them more alarming. All at once I felt like asking him which was the most valuable antique clock, and then smashing it.

"You should be ashamed," he said sternly.

"I should be?"

"Yes. You know how many people never make it to thirty?" He shook his head again, and I was surprised to see that he was suddenly angry. "I had a daughter . . . my only child . . . she was born with a degenerative bone disease. Grew up in a wheel chair. She died when she was twenty-five. She would have considered turning thirty a gift."

"Oh." He was right. I should be ashamed. And I was. I

didn't exactly drop to my knees and thank God that I was going to be thirty soon, but in a little tiny way I realized that Professor Allen had a real point.

I was so surprised by the revelation that I didn't move or say anything else for a while. I just stood there in the middle of the clock museum, listening to the chorus of TICKs and TOCKs, thinking about my age and how lucky I was to have attained it.

"Instead of complaining about your life, why not *do* something with it?" his voice broke into my thoughts. I looked at him, and saw that he was still resentful. I thought about how horrible it would be to have your daughter . . . your only child . . . die so early, and then meet a perfectly healthy woman who was all pissed off that she was getting old. I felt so guilty at that moment that I thought I was going to cry. "I mean, so what if you're going to turn thirty! If you don't mind my saying so, you still have a *lot* of growing up to do."

I dropped my gaze. I wanted to crawl away and not have to face him anymore. But suddenly his voice became gentle and he said, "Forgive me for speaking harshly, my dear. But it upsets me that you don't realize how abundantly you've been blessed."

"I know, I know," I said, nodding earnestly, but still not looking at him.

He didn't respond for such a long time that I finally sneaked a peek at him. I saw that he was giving me his warm smile. "I know I've been blessed," I said. And for the first time, I really *did* know it. I had no right to complain.

He nodded and patted my shoulder. The gesture reminded me of my father, and my eyes filled with tears. "I have clocks to repair, Lark," he said kindly. "You go along now. And think about what I said."

"I will," I promised.

"Good girl."

Awkwardly, he reached out and hugged me. I hurried out. By now I was actually crying a little, and when I got back into my car, I dug around in my purse for a Kleenex. But as soon as I found one, I didn't feel like crying anymore. So I just blotted my nose and said brightly to my fuchsia, "Well! Let's go see if we can't find that Painted Bunting!"

I was blowing down Route 81 when a bright flash of blue caught my eye. Might have been an Indigo Bunting, but I wasn't sure. I immediately pulled into the breakdown lane and got out, squinting at the sky. The interesting thing about the Indigo Bunting is that it isn't really indigo at all, but black. It's the diffraction of light through their feathers that makes them such a stunning blue.

Not a second later, I heard another car pulled over behind me. I sighed. I knew they assumed I was having trouble and they were going to offer help, which was nice, but which was, of course, unnecessary. So I turned to wave my hand and indicate that I was fine, but when my eyes fell upon the car, my hand dropped to my side. It was a beautiful silver limo. Someone with a lot of money was coming to rescue me. I wondered what I was going to tell them.

As I watched, a chauffeur climbed out, circled around the front of the car, and opened the back door on the passenger side. A well-dressed man of about forty emerged. Without acknowledging his chauffeur, he whipped off his sunglasses, tucked them into the neck of his designer polo shirt, and ran a hand through his carefully styled hair, all in one smooth motion.

"Problem?" he inquired, coming right up to me. I saw that he had dark, aloof eyes, and that he smelled expensive.

"Not exactly," I said, and took a stab at explaining about the Indigo Bunting. I figured he would say, "Oh," and leave, but he didn't. He regarded me with curiosity at first, and then with very evident interest.

"You know about birds?" he demanded.

"Some," I said, "Why?"

"What a stroke of luck! My mother just received a parrot as a gift and she doesn't know anything about them. Do you?"

He was awaiting my response so urgently, with eyes no longer aloof, that I felt myself nod authoritatively.

"Sure I do. What do you want to know?"

"Not sure. I wouldn't even know what to ask you. Tell you what," his face lit up suddenly. "How about stopping by the house to take a look? You'll be paid for your time, of course."

"That won't be necessary," I chuckled. Did I look so poverty-stricken that he thought I'd charge him for something like that? "I'll be glad to see the parrot, free of charge."

In the meantime, while we'd been talking, a heavyset woman reeking of wealth, climbed out of the car and scurried over to us. I was impressed. It wasn't exactly scurrying weather, especially if you were a heavyset woman.

"Mother!" said the man, reaching out to take her hand and draw her close. "This girl knows about birds! I've asked her to take a look at Goliath!"

"And what did she say?" The woman was so out of breath that I couldn't help feeling sorry for her.

"She said she would."

"Splendid!"

She beamed at me. I saw that the diamond ring sparkling on her pinkie would have paid my rent for a year.

"Lark DePaolo," I said, extending my hand. She introduced

herself as Hattie Coleridge, and with a very soft, very white, very moist palm, shook my hand.

"Geoffrey Coleridge," said her son, and his handshake was very firm, the way you're supposed to address a tennis racket. He didn't say, "That's Geoffrey with a G," I just knew that was how he spelled it.

"If you would really be so kind as to follow us, my dear," Hattie said regally, "We'll drive you to our home."

"Well, okay," I said. I couldn't help smiling at her. She was a pretty woman, with a delicate film of sweat covering her face, long red nails, soft, white, over-permed hair, and a couple of chins.

"We'll tell André to drive slowly," Geoffrey announced, nodding and leading his mother back toward the car.

"Yes, tell André to . . ." my voice faded off as I watched them climb into the car. "Okay, well I'll see you there," I concluded to myself, then got back into my blue Grand Am. The Indigo Bunting was forgotten in the intrigue of my next adventure. I turned to the fuchsia to explain where we were headed, but it seemed to be saying, "Where's my dirt, Goddammit!" so I didn't tell it anything. I just pulled out after the silver stretch limo and headed down Route 81.

CHAPTER FOURTEEN

We left the highway and travelled for quite a while down long, hidden, windy roads, and then we arrived at a mansion. I'm not kidding, it was huge. It reminded me of Tara, the plantation in *Gone With The Wind*. Acres of lush green property spread out on either side of the house, and in the distance I saw horses. Horses!

I pulled into the driveway, which curved along in front of the house, and watched André hustle out to open the door for Geoffrey and Hattie. As I was climbing out of my car, Geoffrey said, "Leave the key in the ignition. André will move your car out back."

I nodded and did as he suggested. He put his arm out for his mother to hold, and gently escorted her up the front walk. I couldn't wait to get inside the house; I knew it was going to be air-conditioned. Then I glanced down at my poor fuchsia, sitting dejectedly in its two and a half inches of soil.

"André is going to move the car out back," I said. "See you later."

I got out and joined them on the porch, which had patio furniture with matching umbrellas. An elegant man had the door open before we reached it. And inside we were met by gloriously cool air. With a rapturous sigh, I ran my fingers as well as I could through my snarly hair, lifting if off my neck and then shaking it to create a breeze. I could stand to live like this, I thought.

"Over here," directed Geoffrey, heading down the hall. His mother and I followed. She was smiling at me doubtfully as if to say, "It's alright if you don't know what to say about the parrot — neither do I." I returned the smile confidently.

Suddenly a harsh voice said, "Hello!"

At first I thought it must be a very old woman, perhaps Geoffrey's grandmother. But the voice repeated, "Hello!" and I realized it must be the parrot. Goliath. And sure enough, we entered a room in the middle of which, perched with solemn majesty on a hanging bar, was a brilliant emerald green Yellow-headed Parrot. A real stunner. I couldn't conceal a gasp of admiration.

"Hello!"

"Hey, Goliath." I went right over to him with finger eagerly extended. I think I envisioned having him sit comfortably on my shoulder while I discoursed about his lifestyle.

However, Geoffrey said at once, "I wouldn't do that, Lark. Goliath is kind of hostile. The only one he really likes is my brother, Brian."

"Oh." I retracted my finger. A hostile parrot could snap a knuckle in two with his powerful bill.

"So, what can you tell us about him?" Hattie asked. She was standing a distance away, and her hands were gripping one another. Goliath's presence evidently made her nervous. She'd probably never seen a live parrot in her life.

"Fuck you!" Goliath cried abruptly.

"Goliath! You stop that this instant! Oh dear, I wish Brian

wouldn't teach him such vulgar expressions," she sighed sorrowfully.

"Well," I swallowed a giggle, "what would you like to know?"

"I think Mother would like to know if Goliath will die soon," Geoffrey said.

"Not unless he gets sick. Parrots have very long life spans. As a matter of fact, he'll probably outlive you both."

"How lovely," Hattie said grimly.

"Who gave him to you?" I queried. Imagine getting a parrot as a gift? I'd seen them for sale in pet stores, and they weren't cheap.

"My brother Reginald," Hattie answered. "I saw a picture of a parrot in a magazine and said that it had beautiful plumage. The next thing I knew, one was delivered to the door!"

"Christ!" I said, "I mean, wowie."

"And so now we're stuck with it. Oh dear! I just don't know *what* to do!"

"Can't you just explain to your brother that you don't want it, and . . ."

Geoffrey's sudden, urgent shudder startled me into silence. He said in a low, confidential voice, "No, that isn't possible. We have to keep the bird, because of Brian."

"Oh."

"Brian loves that bird."

"Oh."

"It's the *only* thing Brian loves," Hattie added mournfully.

They watched my reaction expectantly, as if they were positive I understood. I said, "Oh," again, as if I did, but I didn't.

"As you can see, we're in, er, a bit of a pickle," Geoffrey said.

"A bit of a pickle, yes, I can see that," I murmured. By now I was really curious about this Brian who seemed to rule their lives in such a loveless way. I wondered if I would get to meet him. "Well," I said, "Parrots make good pets because they're usually affectionate, and because they're content to sit on their perch and not fly around. And as you've undoubtedly noticed, Goliath's pellets are dry and compact, and easy to clean up."

"Yeah, nothing beats that easy-to-clean-up-bird shit," said a voice drenched in sarcasm. We all turned around, and I knew even before I was told that the voice belonged to Brian. I saw that he was overweight, not in a big, sturdy way, but he looked as if he'd never exercised in his whole life. He wasn't huge, just real flabby. He wasn't wearing a shirt, and his chest was white and nearly hairless. He had a dark brown beard and moustache, both untrimmed, and his hair was long and straight and parted at the side, emphasizing his very round head. His eyes looked intelligent, but mean. In fact, that was the overall impression he gave — intelligent, but mean.

"Brian, this is Lark DePaolo. Lark, this is my younger brother, Brian."

"Hi, Brian," I said.

"Hi, Brian!" Goliath said.

Brian didn't respond to either greeting. He was staring at me, not openly admiring me, but startled to see someone who looked like me. He was probably used to slim, vain, cultured women who smoked too much and called him *Brian, darling.* Standing there in my torn hooded sweatshirt, my shorts, my Nikes, no makeup, and that hopeless tangle of hair, I probably didn't look like anyone else he'd ever met.

"Nice to meet you," he finally admitted. I saw Geoffrey and his mother swap astonished glances, and wondered why.

"Where did you find her," he asked them.

"On the highway. She pulled over, and Mother suggested I stop, too, and make certain she was alright."

"And was she?"

"Oh yes. She was just looking for a . . . what was it?"

"An Indigo Bunting."

"Yes, that's right, an Indigo Bunting."

"Of course it may not have been an Indigo Bunting at all," I interjected as if it made any difference to anyone in the room but me. Geoffrey and Hattie nodded politely. Brian just continued to stare at me. An uncomfortable silence ensued. I met Brian's eyes confidently, and he ducked his head with a vulnerable shyness I found appealing, since it was probably very uncharacteristic. "What I saw may have been a Blue Grosbeak. I didn't get a real good look at it, and at a glance, the two are similar."

"Isn't that interesting. And what would the difference between the two birds be, my dear?" Hattie asked. I knew she didn't care, but I was flattered by the attentive lift of her eyebrows.

"Well, the grosbeak is larger and heavier, with a bluish bill. And it has the chestnut wing bars that the bunting lacks."

"Fuck you!" Goliath shouted, making us all jump.

"Honestly, Brian," Hattie said without looking at him, "I wish you wouldn't teach that bird phrases from your appalling vocabulary. It's one thing to hear you say things like that, but to hear Goliath say them when we have company . . . it's just too much."

"I'm sure Lark has heard that word before," Brian observed with a mild smile. I couldn't resist smiling back. Hattie sent me a betrayed scowl, and I tried to frown with her, but couldn't.

"Goliath doesn't have any idea what he's saying, you know," I told her. "Those are just sounds to him."

"Maybe Brian could teach him some different *sounds*, then," Geoffrey suggested with a snotty glare at his younger brother.

"Sure, I know plenty of four-letter sounds," Brian grinned. I didn't want to encourage him, but he was so mischievous that I couldn't help giggling. He was probably only about five years younger than Geoffrey, but the gap seemed greater. Geoffrey could have been Brian's father; he could have been Hattie's brother, not her son. Anyway, Brian giggled, too, and suddenly a kind of bond was formed. I could tell that Brian liked me, and he apparently didn't like many people, so I couldn't help being pleased.

"Will he sit on your shoulder?" I asked Brian.

"Sure. C'mere, Goliath." Brian extended his hand, and Goliath willingly left his perch to travel up Brian's arm. Once settled, he produced a series of contented clicking sounds. Brian grinned at me.

"He's a beaut. I wish I could pat him. But Geoffrey says he's unfriendly."

"Bullshit. If anyone is unfriendly, it's Geoffrey. Goliath won't hurt you. Here — put out your hand."

Timidly, but without hesitating, I raised my arm. Brian took my hand and put it on his shoulder, at Goliath's feet. Goliath glanced placidly at my hand, but didn't move.

"Fuck you," he said.

"Goliath, climb on Lark's arm," Brian instructed. Goliath lifted one foot and nibbled at his toes, then resumed his position, clucking and looking at my hand. Brian reached up and nudged him. Finally he stepped off Brian's shoulder, grasping my fingers and pinching my arm through my sweatshirt as he leisurely made his way to my shoulder. Out of the corner of my eye, I could see his enormous bill just inches away. It wouldn't take any effort on Goliath's part at all to lean over and peck my eye right out. I felt myself getting nervous, and broke into a round of silly, girlish giggles. Enchanted by the sound, Goliath tipped his head and tried to mimic me. Brian was delighted.

"See? He likes you!"

We sent triumphant looks to Geoffrey. Before he could respond, however, another male member of the household entered.

"Hello, all! What's going on? Who's this?"

Hattie's face lit up with a very sweet smile, which made her suddenly look very young and charming. She hurried up to the newcomer, linked her arm in his, and led him to me.

"Darling, this is Lark. She knows all about birds. We've been talking to her about Goliath. Lark, this is my husband, Howard Coleridge."

"Glad to meet you!" Howard boomed, extending his hand with a big grin. His face was very tanned and his teeth looked very white. I shook his hand stiffly, since I had a huge bird sitting on my shoulder. In contrast, his grasp was firm as a salesman's.

"Glad to meet you, too," I said.

"Well! I see you've made friends with the monster! What do you think of him?"

"I think he's really amazing."

"He sure is. You want him? HA HA HA! Just kidding. Where did you meet her, honey?" he asked Hattie.

"She was looking at a . . ."

"An Indigo Bunting," I said.

"Although it's possible what she saw was a Blue Grosbeak," Brian added, "At a quick glance, the two are similar."

"Isn't that something. Well! Who'd like a drink before dinner?"

He clapped his hands together, then rubbed them briskly. We all agreed a drink would really hit the spot. Brian gently relieved me of Goliath and returned him to his perch. As we followed the others out, we heard him call, "Fuck you!"

"Maybe Geoffrey is right, maybe it's time you taught Goliath something new," I said.

Brian shrugged and told me that he considered "Fuck you" one of the most useful phrases in the English language.

"That may be so," I agreed, "but you want Goliath to be well-rounded, don't you? Polite, when the need arises?"

"Fuck you," said Brian. But he was smiling, and I couldn't help smiling back as we followed the others into the dining room.

CHAPTER FIFTEEN

"What's your poison?" Howard began with me. He was still rubbing his hands together in heady anticipation, and grinning at me like I was a daughter he hadn't seen in years.

I wanted to request a mint julep because it had such a crisp, refreshing sound to it, but I've always heard they're really brutal tasting, so I told him I'd like just a glass of seltzer water.

"I'll have seltzer, too," Brian said. Howard stared at him in amazement.

"Since when do you . . ."

"I just feel like it, okay?" Brian interrupted, suddenly sullen and humiliated.

"Oh, you have a pool," I changed the subject hastily. I was looking out the window past Howard's head at a shimmering, rectangular, built-in pool. After being in a hot sweatshirt two steaming days in a row, to dangle my feet in that clear blue water would be a slice of heaven.

"Sure do!" Howard said, "Like to go for a swim?"

"I'd love to," I answered happily, but then remembered that I didn't have a suit, and shook my head. "Can't," I said. "All I have is this." Irritably, I tugged at my blue hooded sweatshirt.

"I'm sure we can find something to fit you, my dear," Hattie said at once. "My daughter is just about your size, and when she got married and moved out last year, she left a couple of bathing suits here."

"Wow, that would be great."

"Yeah, a swim would really feel great," Brian agreed.

Meanwhile, Howard, Hattie, and Geoffrey were sending each other looks that demanded, "What's gotten into Brian?" And actually, it was a little frightening, the way he couldn't keep those intense brown eyes off me. Not that I thought he was madly in love with me, or anything . . . but I could tell that, for some reason, he was attracted to me. Every time I glanced over at him, he was watching me; and each time our eyes met, he'd drop his gaze, only to look up again as soon as I turned to someone else.

Howard served the drinks. He didn't ask Hattie or Geoffrey what they wanted, he was obviously familiar with their preferences. Theirs were pink and gold respectively, sparkling with ice cubes that tinkled and glistened. I almost changed my mind and asked for some of whatever they were drinking, but I remembered in time that I always like the look of alcohol more than the taste. Besides, it's fattening.

Exchanging some small talk, we wandered out to the pool. Without waiting for an invitation, I kicked off my pink Nikes and stood on the top step, ankle deep in water. It was the perfect temperature — not heart-stoppingly cold, but cool enough to take the edge off the heat. Brian, who was already barefoot, stepped down next to me.

"Feels good," he said with a shy grin.

I nodded rapturously, and took another step, to my knees, then another, until I was up to my thighs in that glorious water.

"Why don't we see what we've got to fit you?" Hattie

suggested from behind.

I didn't answer right away. I'd shut my eyes, and could feel the hot sun making my face glow. All around me was silence, which I suddenly realized was the sound of everyone eagerly awaiting my response. Opening my eyes and turning to look at them all, it struck me as being kind of funny. Not amusing. Odd.

"Well, okay," I said, reluctantly climbing out of the water.

"Splendid," Hattie said, handing me a towel she'd picked up off a chair. "Dry your legs and follow me."

We went back inside, and she led me through several rooms, down a long hallway, and then up a wide, richly carpeted staircase. The house was so huge that I thought I could stay a year and still not get a chance to see it all.

"You have a lovely home," I remarked graciously, momentarily sounding artificial, like my mother. *"You have a lovely home"* was definitely something she would say, even if she was visiting convicts in a prison.

"Well, it isn't much," Hattie protested, and began to speak at length about the problems they'd had with the servants when they first moved in. As I half-listened to her prattle, I looked around at everything in envious astonishment. Expensive-looking paintings covered the walls, exotic knick knacks covered every gleaming table and shelf. I've never been a money-oriented person, but it was all I could do to keep from saying, "How much did this cost? How much did that cost?"

"And this is Trixie's old room," Hattie said, stopping so abruptly that I smacked into her and had to apologize. Up close, she smelled very powdery. I looked beyond her into Trixie's old room, and saw that it was frilly and pink. The bed even had a canopy. As I studied it, it occurred to me that I'd never seen a real canopy, just pictures of them in books of fairy

111

tales. Hattie preceded me in, disappeared into a closet that was bigger than my mother's dining room, and emerged a moment later with three string bikinis.

"One of these should fit," she declared. "I'll leave you to try them on. You and Brian can go swimming together. Won't that be nice?"

I nodded, staring in dismay at the three tiny bathing suits. She left, gently shutting the door behind her. Didn't Trixie own any one-pieces? Maybe she'd taken them with her when she moved out. I selected one that I thought was a teensy bit bigger than the others, put it on, then apprehensively checked myself out in a mirror which ran the entire length of one wall. I've always been self-conscious of my extra pounds, almost neurotic. I'm one of those women who practically have to be on fire before she'll appear in public in a bathing suit.

But suddenly something Alex once said came back to me, something I hadn't put much stock in at the time. He told me that women are too hung up on their bodies. "You think that all men care about are big breasts and flat stomachs," he said, "but the truth is, that's all *women* think about. In bed, men don't care how a woman is built. They're just grateful to be in bed with her." Besides, he told me, I had nothing to be ashamed of, I had a very nice body. "You're young and you don't smoke or drink, and you're full of energy. And I mean, what could be more beautiful than good health?"

Examining my reflection, I realized that he'd been right. Instead of agonizing over my little round belly, I should be thankful that I wasn't sick or diseased. Professor Allen's advice came back to me, too, about how I should stop complaining and start being more grateful. And all of a sudden, I felt okay about myself. What a strange feeling it was! Strange and wonderful. I heard myself laugh. It had been an amazing day.

Recklessly, I hurried back down the stairs. But once I reached the hallway, I couldn't remember which way I was supposed to go to get back to the pool. I've always had a terrible sense of direction, inherited from my father. My mother was the one who was shrewd with directions, and always discovered short cuts that flabbergasted my father and me.

I went to the right, and entered the first room I came to, where I startled the butler. He started to ask who I was, but I was out the door before he finished. I came to another room, but just before I went in, I heard voices. Hattie, Geoffrey, and Howard. Impulsively, I stepped back and pressed myself against the wall and kept quiet, listening to them.

". . . really seems to like her. I've never seen him like this! He's being so friendly!"

That was Geoffrey. I heard ice cubes going CLINK CLINK. Then Howard said, "You know, she might be good for him. Maybe we can keep her here for a little while."

"Make her feel so welcome that she won't want to leave," Hattie put in.

"But what about . . . I mean, she must have some place to go," Geoffrey pointed out pragmatically. "How could we possibly talk her into staying with us?"

"She has an out-of-state licence plate. She must be on vacation," Hattie said.

"Well she can vacation with us for a few days," Howard said heartily.

"But what will we tell her?" That was Geoffrey again. I could tell he was frowning, and I pictured him tapping his glass with one of his long, slender fingers.

"We'll just say . . . we'll just say that we love having company, but that most of our friends are away."

"I really *do* love having company," Hattie said.

"And she might be good for Brian," Howard said again.

Well by now I was *real* curious to know what was going on. It was eerie, listening to them make plans to keep me hostage.

"Let's go ask her," I heard Howard say, and in a flash I ran back down the hall, and miraculously found my way to the pool.

Brian was already in, up to his chest. When he saw me in that tiny bikini, his eyes opened wide.

"Thank God my sister had a penchant for small bathing suits," he said appreciatively. I acknowledged the compliment with a modest smile, and stepped into the water and headed over to him, until I was standing right next to him.

"You're so lucky to have a pool. Do you live here with your parents?"

"Wouldn't you?" he asked with a sardonic grin. "Got everything I need. Butler. Chef. Maid. Nice car. Besides, it's been generally accepted that I'm not competent enough to live alone."

"Not competent enough?" I echoed, baffled. "What does that mean?"

"It means I'm psychotic."

I didn't respond at once. I didn't know him well enough to be able to tell if he was kidding. I pinched the water with my hands so that it squirted. I noticed that it was way past time to remove my pink nail polish.

"Are you really?" I asked finally.

He shrugged, then nodded.

"That's the what the experts are saying."

"But . . . well . . . how do you, I mean, what kind of symptoms do you have?"

"Well, every once in a while I get kind of crazy . . . kind of violent." He tried to sound casual, but it obviously was painful for him to speak of it at all. I admired him. It was a major confession. I doubted *I* would go around telling strangers I was psychotic.

"I'm sorry to hear that," seemed to be the best thing to say. Brian chuckled, then shrugged again.

"Oh well," he said.

"You seem pretty normal to me," I offered. "I mean," I added, seeing I'd offended him, "you seem pretty well adjusted. Maybe more intelligent and intense than normal."

He liked that better and smiled and nodded. Then he leaned close and whispered, "They're watching us, you know."

Uh oh, I thought, here we go. Who's watching us? Aliens? FBI?

"Look up there."

With his head, he motioned toward the house. I turned, and sure enough, I saw a curtain move at one of the windows, as if someone had hastily stepped out of sight.

"Oh, you mean your brother and your parents?" When he nodded, I said, "Why?"

"I think they want to make sure I don't attack you. Believe me, I'd like to." The smile he gave me was so sexy that I actually tingled a bit, even though I wasn't really turned on or anything. "And I think they're curious about the way I've been treating you."

"The way you've been treating me?" I felt like everyone understood something I didn't have a clue about.

"Well," he said, his eyes leaving my face and wandering to some point beyond me. Moving his hands across the water, from left to right, from right to left, twisting at the waist, he went on, "I don't know. Usually I'm pretty aloof or scornful or whatever you want to call it. But with you it's different. I mean, I know we just met, but I really feel comfortable with you."

115

I almost laughed. It's a line every woman in America has heard at least three times. But Brian said it so humbly. It sounded sincere.

"Well . . . thank you," I mumbled awkwardly.

"I mean, I'm not saying I've fallen in love with you or anything. I just don't hate you."

"How sweet. I don't hate you, either," I said. We both smiled.

Just then, the other members of the family appeared. And for the rest of the afternoon we chatted about this and that, and drank our drinks, signaling for a servant when we needed a refill. Brian and I swam in the pool, and all of us enjoyed the warmth of the sun. Hattie invited me to stay for dinner, and I agreed happily. I was having so much fun that I forgot that was exactly what they'd intended.

CHAPTER SIXTEEN

Hattie excused herself, and reappeared several minutes later, wearing a mint green evening dress with long sleeves and a very full skirt. She explained that she always liked to dress for dinner, particularly if she was entertaining. For a moment I assumed she meant that guests were arriving; but then I realized she was referring to me. I wondered if she expected me to change, too. Reluctantly, I sat up in my lawn chair. But to my relief, she was smiling warmly at me, without any expectations at all.

"Dinner will be served momentarily," she said, "if you'd like to get dried off." With one hand, she indicated a stack of folded towels, which looked thick and new, and which lacked the faded cartoon characters that characterized the beach towels I had at home. They were even the same color and everything. "I'll see you inside." With a nod, she turned and left again. She'd cultivated a strikingly graceful manner, which I couldn't help admiring.

"I really love your mother," I confided to Brian as we got up and headed for the stack of towels. But he just shrugged.

"All set?" Howard inquired after Brian and I had dried ourselves off. I nodded. "After you, my dear." Graciously, he stepped aside, allowing me to precede them all inside. But I hung back, unsure of which way to go. With a chuckle, Geoffrey took the lead, and a moment later we entered a huge dining room, not the one we'd had drinks in earlier, but one with a table as long as a Winnebego. Comfortably clustered at one end, Howard sat at the head, with me seated next to Brian, and Hattie between her husband and Geoffrey. Apparently Geoffrey lived at the house, too. I wasn't sure why. It was difficult to believe that two sons would have been deemed unable to live alone, but at that point, nothing would surprise me.

A couple of maids served us dinner, and it sure beat the hell out of Luigi's pizza. We began with French onion soup. It was so rich that as soon as I finished, I was stuffed and ready to leave the table. But then salad arrived, full of exotic ingredients it would never occur to me to buy, like raisins and avocados. So I ate that, too. Then we were served the main course, which I assumed was chicken until Geoffrey said, "Mmm, what tasty pheasant." I said smoothly that it was the best I'd had in years. And then the maid brought in cake. Cake! And it wasn't even anybody's birthday! By that time I was so full that I could only handle a small piece. It occurred to me that I'd probably regained in a single meal all the weight I'd lost by starving myself for the past several days, but I didn't really care. It was high time I stopped worrying about a couple of extra pounds.

Hattie asked if I'd like coffee, but I had to shake my head. I felt like a Turkey Vulture must feel when he's eaten so much he's unable to fly. All the food had made me lethargic and drowsy, and I wanted nothing more than to take a nap.

There was a bell on the table, and as soon as I said No to coffee, Hattie reached over and rang it. Instantly a couple of maids appeared and began to clear the table. I was fascinated.

Imagine living like that? I thought about dinner at my house when I was growing up. We'd sit down, then one of us would remember we needed the butter. The other two would put down their forks, while the one who wanted the butter would leap up and retrieve it. We'd all settle down again, but then someone else would realize that we needed a serving spoon for the vegetables. Once again, we'd put down our forks and wait while the spoon was fetched. And so on, for about ten minutes each evening.

After dinner Brian asked if I'd like to sit in the jacuzzi. I'd never been in one before, but it sounded tempting. I was still wearing that tiny bikini, and inside the air conditioner was blasting. All during dinner I'd shivered, and my fingers had been so cold I'd barely been able to hold my fork. So when I lowered my frigid body into that steaming water, I thought I would weep with pleasure.

"Who says money can't buy happiness," Brian chuckled, climbing in next to me.

"Mmm," I sighed, leaning back and shutting my eyes.

"So you'll spend the night?" Hattie's eagerly shrill voice startled me out of my intense state of relaxation. My eyes shot open — I hadn't even realized she was there — and I sat up and turned around to stare at her in surprise. Geoffrey and Howard were there, too, regarding me hopefully..

"Spend the night?"

"Unless you've already made reservations to stay somewhere else. Have you?"

"Like at some shitty, run-down motel?" Brian asked, and the look in his eyes said, "Come on, admit it — you'd love to stay here, you love this place."

I looked from one to the other, amazed that they felt so comfortable inviting a total stranger to spend the night.

"My mother loves company," Geoffrey said.

"And most of our friends have gone away," Howard added.

"Oh yeah, that's right," I murmured, recalling their conversation of several hours ago. They were determined to make me stay. But why? Just because Brian liked me? There had to be more to it than that, didn't there? What would they do if I turned down their offer? Would they go along with that . . . or would they insist? Use force? Despite the heat of the jacuzzi, goose bumps crawled up my skin. Was I in danger? Maybe they were *all* psychotic. Maybe I'd never get out of alive. Maybe I'd have to stay there and be served elegant meals and swim in the pool and relax in the jacuzzi and even learn to ride those horses out back, for the rest of my life, with no hope for escape. Luckily, I had nothing else planned.

"Okay, I'll stay," I said.

Relief shone on everyone's face.

"I'll have André bring in your bags," Hattie beamed, turning to leave.

"Wait," I said awkwardly, "I mean, I don't have any bags. I didn't, um, I didn't bring any. I just, well, it's kind of a long story."

"No need to explain," Geoffrey put up his hand to halt my uncomfortable rambling.

"Then you'll need to borrow more of Trixie's clothes," Hattie said merrily, as if the prospect delighted her. She even clapped her hands together, reminding me of a young schoolgirl.

"I guess so," I said.

"As soon as you get out of the jacuzzi, we'll go up and see what she's got that appeals to you."

"Trixie's clothes won't fit Lark," Brian objected. "Trixie is fat. Lark is thin."

I stared at him, flattered, but convinced that everyone was

right, that he *was* psychotic.

"Trixie isn't fat," Hattie corrected firmly. "She's voluptuous."

"Like her mother," Howard put in. He and Hattie swapped amorous glances, which set them both on fire. At the same time, they both reached out their hands, and their fingers met and clasped. It was so intimate that I felt embarrassed for having witnessed it, and looked away. It was the kind of gesture you'd see take place between newlyweds, not a couple who'd been married for years and years.

"Whatever," said Brian. "But that suit was always too small for her. It fits Lark perfectly, though. You might as well keep it," he told me.

"Of course she can keep it," Hattie interjected generously, "that is, if she *wants* it. I mean, it's last year's style."

They all turned to look at me with expectant half smiles.

"I knew that," I assured them all, "I knew it was last year's style."

"You can keep anything you find that you like," Hattie went on. "Whatever Trixie left here, she no longer wants."

"She has too many clothes, anyway," Brian said.

"Nonsense, dear, no woman has too many clothes. That doesn't even make sense," Hattie scoffed.

"Well . . . if you're sure . . . that would really be nice of you . . ." I felt like a welfare mother, accepting everything they offered me. "Thank you." I started to rise, but Brian wrapped a hand around my shin.

"Don't go yet. You have plenty of time. Just stay here for a little while longer. Okay?"

"Okay." I sat back down, feeling absurdly pampered and loved and comfortable.

The others discreetly excused themselves and left, and once

121

again, Brian and I were alone. There was a bit of silence, but it didn't bother us; neither felt the need to fill it with senseless small talk. But then I asked Brian what it was like to be psychotic, and abruptly he began to tell me about himself.

"When I was very young," he said, and his voice was gentle and sad, "I never had any friends. I was always different from other kids my age. I was always depressed. I never laughed or played with them, I just sat around and thought about how shitty my life was."

"How come?"

"Because it was."

"Why?"

"Because! Because I was very sensitive. Because I saw and understood things most people aren't ever aware of."

"Like what?"

"Like despair, that's what."

I frowned. I felt like he wasn't really giving me any real answers, and I didn't know if I should keep probing him or not. He looked like he was disappointed that I didn't know exactly what he was talking about, and I suppose it would have been easier to pretend that I did, but I didn't feel like it. I wasn't even sure *he* knew what he was talking about.

"Lots of people know about despair," I said.

"Not the way I did," he said. "For years, I tried to devise ways to kill myself."

Nausea shot through me and I sat up so suddenly that he was startled. "Don't say that," I said.

"It's true. I've always been obsessed with suicide. When I'd get real depressed, I'd write down different methods, to cheer myself up. Of course I considered buying a gun, but that's so trite. I wanted a more creative way to kill myself. I considered hanging myself, but . . ."

"Stop it, stop it!" I cried, covering my ears. "Don't talk about it anymore!"

Brian was surprised by my distress, and quieted right down. But then the silence grew unbearable, and he had to continue. "That's why I can never live alone — because everyone knows I'll kill myself. I have people watching me all the time. Hell, I don't even know why they bother. My family doesn't love me. Why should they care if I live or die?"

I started to say, "Of course they love you," but I didn't think he was expecting me to, and besides, I wasn't completely sure it was true. So I just nodded.

"Growing up, I always felt alienated from them. Geoffrey and Trixie, who are both older than me, were real popular in school. Trixie was always beautiful and fashionable, and Geoffrey always excelled in sports."

"Geoffrey?" I echoed, amazed. I thought back on the jocks at my high school. No matter how hard I tried, I couldn't quite picture Geoffrey shooting baskets or hitting home runs or scoring touchdowns.

"Oh yes, he's an excellent golfer, and his polo game isn't too shabby, either."

"Oh."

"So I became more and more withdrawn. I started spending all my time reading. Sartre. Nietzsche. Kafka."

"Huh."

"No wonder I grew up so arrogant."

"Did you go to college?"

"I tried. I had great hopes for college. I thought I would find some kindred souls or something. But I didn't. I was still different, and they were still all the same. I got involved with drugs and alcohol. I almost overdosed a couple of times. Finally my parents and my therapist and my parents' therapist

123

decided I'd be better off at home. So I came home, and I've been here ever since."

"What do you do all day?"

"I still read. I write, too. Poetry. Cynical, tortured poetry."

"Oh." I didn't ask to read any. Poetry had always bored or baffled me. I knew instinctively that Brian's prose would be saturated in esoteric symbolism, and I was afraid that if I didn't like it or understand it, I'd see his violent side, and he'd kill me. So I just sat there in the jacuzzi, nodding and not asking to read it.

"I've written dozens of poems — enough to fill a book. I've been thinking about submitting it to a publisher. Don't ask me why," for a minute, there was an almost hopeful look in his eyes, "but I feel like . . . deep down inside . . . if I can use my unhappiness to accomplish something, then I won't be haunted by it anymore. Do you know what I mean?"

"Sure. All your life your attitude has been destructive. But if you can begin to use it in a constructive way, it will take on a whole new meaning for you."

"That's exactly what I was trying to say," he said, and his smile was warm and sincere. It occurred to me that he could have been good-looking if he smiled more often.

Just then Hattie appeared.

"Still in the jacuzzi?" she asked, glancing at her watch. I realized it was getting kind of late, and that she'd been waiting for me to go look through Trixie's clothes. So I told Brian I'd see him later, and reluctantly stood.

"Here's a towel for you, my dear," Hattie said, handing me one. "Dry off and we'll go."

I rubbed myself briskly, then wrapped the towel around my chest, inhaled deeply, and tucked the corner in under my armpit.

But the second I let out my air, the towel fell to the ground. I was humiliated and annoyed. I've always wanted to be one of those women who can wear a towel like that, but it just doesn't stay up. Sheepishly, I threw it around my shoulders instead, and followed Hattie back up to Trixie's room.

"Now let's see," Hattie's voice faded as she disappeared into the depth of the enormous closet. I went in too, looking around in amazement. There were *dressers*! In the *closet*!

"This is awfully nice of you," I managed to say. I couldn't picture myself in any of the clothes I was seeing — probably one of Trixie's dresses cost more than all the clothes I'd ever purchased in my whole life.

"My pleasure," Hattie beamed. "Having you around is like having my daughter around again . . . only better, because you're so easygoing. Trixie tended to be a little, well, difficult. But of course I loved her. I love *all* my children." She was gazing at me with a significant gleam in her eye that I didn't understand.

"Of course," I said.

"I'll leave you alone now, my dear. Help yourself to anything. Is there anything you need?"

I assured her I'd be fine, and we swapped Goodnights. Then, with a motherly smile that made me a little homesick, she sailed out.

CHAPTER SEVENTEEN

I peeled off the damp suit, rinsed it out, and threw it over the towel bar in Trixie's bathroom. Then I began to dig through some of the dresser drawers, and came upon the prettiest linen nightgown I'd ever seen — white, sleeveless, with delicate lace around at the collar. See-through in the right light. As I modeled it in front of the mirror, I felt like a million bucks. I felt a little funny not wearing underwear, but I didn't feel close enough to Trixie yet to share panties with her. Besides, going without seemed to be appropriate to my new lifestyle. I felt wild and reckless, like I could do anything I wanted to. As I snooped around Trixie's room some more, I realized I was humming what little I could recall of what I suspected was the new Food tune. I felt happy and excited, in a way I hadn't in years and years.

A knock on the door startled me.

"Come in!" I called out gaily.

I was surprised when Howard poked his head in. His eyes lit up when he saw what I was wearing.

"I remember that nightgown," he said, "but it's been a long time since I've seen it. When Trixie was younger, she used to wear it all the time."

"It's really pretty," I said, fingering the lace at the collar which wasn't even itchy.

A sentimental smile crossed Howard face. He said, "Sometimes when she wore that, she'd come into my office to say goodnight, and she'd sit on my desk, and talk to me about her boyfriend problems. Seems none of them measured up to her old dad," he chuckled sheepishly and with poorly disguised pride.

I said that was sweet, but he was so caught up in the memory that he didn't hear me. I thought of my own dear father, and at that moment, I liked Howard very much. A second passed during which neither of us spoke, and then he cleared his throat.

"Ask you a favor?"

"Sure," I said. "What."

"Well, it's about Brian."

"Yes?" Suddenly I felt *real* funny. I remembered the strange conversation I'd overheard earlier, and couldn't wait to find out what was going on.

"Brian is, well, he's sort of . . . he's kind of . . ."

"Psychotic?" I suggested.

Howard's mouth dropped open. Then he nodded. "Psychotic," he affirmed reluctantly. "Is it that obvious?"

"No, he told me."

"He told you? He told you he was psychotic?"

I shrugged and said, "Yeah."

"Well, that means he trusts you. That's good."

"Why."

"Well, because, um . . . I was thinking, I mean, *we* were thinking . . . Geoffrey and Hattie and I . . . we were thinking . . . that Brian really seems to like you . . ."

He waited for me to comment on that. But I couldn't come

127

up with anything to say, so I just watched him and waited for him to go on.

"I was wondering . . . I mean, *we* were wondering . . . if you would please do us a very big favor."

"What," I said warily.

"Well, it's just that Brian doesn't like and trust very many people. And he really seems to like you. You see, all his life he's been unhappy and . . ."

"WHAT!" I interrupted because I couldn't stand not knowing what the favor was for one more second.

"Lark, I'm asking you if you'll sleep with Brian. You can say No if you want to."

My eyes grew wide, and I dropped in astonishment onto the bed, staring at an acutely uncomfortable Howard.

"Sleep with Brian?" I repeated, in case I hadn't heard him correctly. Maybe he'd asked me to *sweep* with Brian . . . or *leap* with Brian.

"You may find this hard to believe, my dear, but Brian has never, um, *known* a woman."

"I do find that hard to believe," I agreed. "How do you happen to know that?"

"Well, since he was diagnosed we've kept a very close watch over him. He's gone on dates, but there's something kind of, I don't know, frightening about him, that the women don't seem to go for. And he's put on a lot of weight over the past couple of years. He never leaves the house anymore. You're probably the first woman who isn't related to him that he's seen in a year or more."

"Holy cow. But . . . why do you want me to sleep with him? I mean, what will that accomplish?"

"I don't know. Maybe nothing. But it seems to me that if he can experience what it's like to be intimate, it might have a positive effect on his attitude. He might get a feeling for what it's like to care about someone besides himself. You know, for a long time Hattie and I considered hiring a, you know, call girl . . . but then we agreed that it might make matters worse. He probably wouldn't develop any romantic feelings for her, and he might become even more cynical."

"Huh," I said, thinking about everything he'd said. It occured to me that I should be offended, I mean, what did he think I was, anyway? But I wasn't offended at all. A week ago I probably would have been. But I couldn't help thinking that Howard was right, that maybe intimacy would help Brian. And I liked Brian and wanted to make him happy if I could. And it wasn't like I'd be cheating on Alex, since we'd split. Finally, I felt myself nod willingly.

"Sure, okay," I said.

"Really?" he was aghast, and I considered being ashamed of myself. But then he smiled, so relieved that I smiled, too.

"That's just *too* generous of you," he said, and he even reached into his pocket, withdrew a handkerchief, and mopped his brow. I wondered if he'd ever sweated before, or if that was the first time. "Now we have to be careful. We don't want him to suspect we've set this up."

"Leave this to me," I said briskly, with a nonchalant flip of my hand. "He won't suspect a thing."

"Well, alright, then. Hattie and I are going to bed. Geoffrey has gone out with some mah-jongg buddies. So it will be just you and Brian. Last I saw him, he was in with Goliath."

"I'll go down," I said.

"Okay. Um. I really appreciate this, Lark. I want you to know that this isn't something I do all the time . . . I mean, I've never done it. It's just that he really . . ."

"Seems to like me."

"Right."

"Well, I like him, too. I want to help him."

"I like him, too. I mean, I love him. I mean, we all love him. I don't think he knows that. You might want to mention that while you're . . ." Horrified at his near faux pas, Howard slammed his mouth shut and fled down the hall. I listened to his footsteps fade, and took a couple of deep breaths. Then I went down the stairs and headed for the room where Goliath was kept.

CHAPTER EIGHTEEN

Brian glanced up with a nervous smile as I entered. His eyes opened wide as they took in the long white nightgown.

"You look just like an angel," he said, startled and bashful. In light of my intentions, the comparison made me blush. I wondered how I was going to lure him to bed.

"Fuck you," Goliath greeted me amicably.

"I thought you were going to teach him something new."

"I did! Listen to this — Goliath, tell Lark your new word."

"Fuck you."

"No, no! The one I just taught you!"

"Fuck you!"

"Hmm, he seems particularly fixated on that one," Brian grinned and shrugged. I grinned, too. There was some silence, and this time we were both sharply aware of it. What if Brian didn't want to sleep with me? Then what? Or what if I had sex with him, and he was still depressed? Or what if he became so attached to me that I had to marry him?

"This is Trixie's nightgown," I said abruptly.

"I know."

Brian was breathing heavily, not out of any exertion or sexual stimulation, but because he was so apprehensive. I would have given anything to know what was going through his mind. He was having trouble meeting my eyes, and seemed overly preoccupied with the newspaper under Goliath's perch. With one toe he kept trying to make it parallel with the tiles on the floor. He was still barefoot, but now he wore a tee shirt which bore a faded, morose Edgar Allen Poe. I could tell he wasn't going to make any overtures. It was up to me. And that was unfortunate, because, I realized all at once, I'd never had to initiate sex with anyone before. They guy had always done that; and if he hadn't, I'd assumed he wasn't interested, and gave up on him. What in the world am I doing? I agonized, now as nervous as Brian.

"So," I said.

"So," he repeated.

"*So!*" Goliath cried, suddenly flapping his wings and startling us both.

"Hey, how come Goliath doesn't fly around the room?" Brian asked briskly, as if new information about his parrot was the most important thing in the world.

"Most parrots tire very easily when they fly — their wings are so large that it's an effort for them. They prefer to sit or use their toes to walk along branches."

"Oh."

"Um, Brian?"

"Yeah?"

I dug through my brain for something to say. I figured that if I could get him to take me to his room, the rest would be easy. Like most American women, I'd been taught that men can't be in a room with a bed with you and not insist on having sex. "How about showing me some of your poetry?"

He was visibly unnerved. For the first time, our eyes met. He looked wary and baffled, like his life had suddenly ceased to make sense. I almost took pity on him. But hell, I had an assignment.

"Well, okay, I'll go get some and bring it down."

"No, I want to see it in your room," I said brazenly. He looked away again, and resumed trying to position the paper beneath Goliath's perch.

"Uh, okay."

Neither of us moved or spoke for several long seconds. And then he turned and fled, without even looking at me. I had to run to keep up with him. We went up the wide, carpeted stairs, and down the hall, past Trixie's room. The door he stopped at was closed and had a *Do Not Disturb!* sign hanging from the knob. He opened it and turned on the light. I followed him in and shut the door behind us. The room was a mess. There were clothes and candy bar wrappers all over the floor, and on the unmade bed were a couple of Stephen King novels.

"Wow," I said.

He didn't answer. He was facing me, but not looking at me. He looked very vulnerable, and I noticed that his fingers were clenching and unclenching at his sides. His chest heaved while his breath ran in and out.

"Brian," I said.

He didn't answer. I reached out and raised his chin with one hand so that he'd have to look me in the eye. He did, for the tiniest second, and I saw that he was scared to death. I felt like a giant predatory hawk who'd just swooped down on an innocent bunny.

"Can I stay with you tonight?" I asked. My God, I thought, I've just propositioned a psychotic.

He took a step back so that he could look away again. It occurred to me that I'd been counting on him being lecherous and eager. This relentless nervousness was tough to deal with.

"If you want to," he said eventually.

"Do you want me to?"

A long time passed before he finally nodded.

I felt myself release a breath I didn't realize I'd been holding. Cool relief spread through me. Weak, I dropped onto the bed. I hoped he would sit next to me so that we could start, but he didn't.

"How about showing me some of that poetry," I reminded him.

Nodding, he picked up a notebook from his desk, leafed through the pages, then handed it to me. I saw that it was opened to a poem titled, "Needles, Bottles, & Bed." Apparently he'd written it during his college years, a period of his life when he was either stoned, drunk, or passed out. I was impressed by how powerful his language was, and his vivid descriptions. He had a gift, that was obvious. But the poem was so desperately bleak that I couldn't even finish it.

"What an awful time for you," I said feebly. "You must be glad it's over."

"Well, sometimes I wonder."

"You wonder what?"

"I wonder if my life is any better now. I mean, back then I was buried in oblivion most of the time. And now, well, I wonder if reality is any great thrill."

I frowned. If he expected me to feel sorry for him, he was going to be disappointed. "Oh quit whining, Brian. 'Buried in oblivion' is no way to live."

"Precisely. That's why I tried to kill myself."

"But your life doesn't have to be oblivion or suicide! Christ! I mean, there are other options!"

"Like what."

"Like changing your attitude. You're such a martyr! Why are you so eager to suffer? You know, it's okay to be happy, Brian. It's not a crime or anything."

"That's fucking naïve," he said, and his tone was defensive.

"No it isn't. You have to stop thinking you're making a point by being miserable despite the fact that you're rich. You think you're so important that your unhappiness makes any difference in the world? It doesn't. You're *tiny*, Brian. You know what your problem is? You take yourself too seriously. You've been given a good life. Why the hell don't you enjoy it?"

I pushed the notebook aside and glared at him. He closed it. I was afraid he was pissed, and I was sorry about that, but I couldn't help it. I was pissed, too. I have no patience for anyone who glorifies despair and dreams of suicide like it's noble or courageous.

"Well, maybe," he mumbled, surprising me.

"Really?"

"Maybe. I mean, I guess you're right. It's just that . . . happiness doesn't come naturally to me. I think that being happy is hard work. Being miserable is so much easier."

"I know," I said, and I was thinking about myself. Who was I to give Brian advice about enjoying his life when it was so good? Didn't I used to have a good life, back home with Alex? A pang of loneliness for him shot through me, and it hurt so much I couldn't catch my breath for a second. For the millionth time, I wondered why I'd been so unhappy lately.

I heard Brian sigh deeply, and looked over at him. I was hoping at that moment to see a man transformed and full of appreciation for his blessed existence, but all I saw was a dark, reluctant scowl. I could tell that he kind of knew my advice was worthwhile, but that he didn't want to admit it or act upon it yet. So instead of shrieking, "Eureka! You're absolutely right!" he suddenly said, "Alright, what about you. What are you doing here all by yourself? I mean, you must have a boyfriend or a husband or something."

"I did," I sighed, and briefly recounted my recent split-up with Alex.

"Tell me more," Brian instructed, and together we stretched out crosswise on his bed. It was a huge bed, with a big puffy comforter that felt nice and soft on our bellies. And before I knew it, I was telling him all about my childhood and my father.

"We used to take these long, spontaneous trips together. I was so happy then. I guess I'll never be that happy again, as long as I live. And there's nothing I can do about it. Those days are gone forever. There's nothing I can do to bring them back. It doesn't seem fair, you know? I mean, to have those wonderful times just snatched away."

I paused and drew a shaky breath. It would have been easy for me to burst into tears, but I didn't want to. I guess Brian didn't want me to, either, because he hastily asked me to tell him about my mother.

"I don't know," I shrugged. "I was never that close to her. I guess I was too little to have enough love in me for both of them. I adored my father, and I'm afraid there wasn't much left over for my mother. Don't get me wrong, I mean, she's a terrific woman, and she was always there for me . . . well, I mean, she would have been there for me, if I'd ever needed her. But I never did. I always turned to my father. And when he died . . ."

I stopped. Tears were flooding my eyes, and I didn't make any effort to stop them from spilling down my cheeks. Brian didn't say anything, just watched me sympathetically. I felt him stroke my hand, and I wondered if he'd ever done that to anyone else before.

"When he died, I wanted to die, too. I tried. I stopped eating. I didn't even get out of bed for two months. I just lay there, not speaking or moving, hoping I would die soon, so that I could be with him again. I didn't want to go on. I didn't think it was possible to live after he was gone."

"What's going on, Mother? Why is there an ambulance out front? Is something wrong with Daddy? Where is he? What's going on?"

"Don't go in there," my mother warned, barring my way and trying to hug me. I could see she'd been crying, and was attempting despite that to be brave, but that didn't matter. I tried to push her aside.

"What do mean, don't go in there? Of course I'm going in there!" I shouted angrily. I was so worried that I would have hit her if I had to, to get her out of my way.

"No, Lark, don't." Her voice was firm. Her grip on my arm tightened. I struggled to get away, but I couldn't. I hadn't known she was so strong. I called her a terrible name, which hurt and startled her, but not enough to let me go.

"You can't go in there because your father . . ." her voice faltered.

"What? What is it? Let go of me! Tell me what's going on!"

Now I was so scared I was crying. I tried to scratch her arms, but she still held me tight.

"He's . . ."

"He's what? He's what?" I tried to get her to look at me, but she wouldn't. I was filled with such horror that I couldn't speak or ask again what had happened to him. I think it was

137

then that I realized he was dead. I let out a shriek that could probably be heard outside. Finally struggling out of my mother's grip, I tore into my father's room.

A couple of paramedics hovered over him. It seemed to me that he was still moving, but maybe that was because they were lifting him onto a stretcher. There was blood all over, the air stank of its thick heaviness. I saw that he'd slit his wrists, vertically along the veins, the way you're supposed to. Just before they hurried out the door with him, headed for the ambulance, I caught sight of his face. It was very white, and his mouth was hanging open. I tried to touch him. I tried to pull him off the stretcher. I heard one of the paramedics say, "Christ! How did she get in here?" and someone grabbed me and thrust me into my mother's arms. She tried to hug me, but I pulled away, screaming, "He killed himself because of you! Because you said you were leaving him! You murdering bitch!"

"I loved him so much!" I wailed, clinging to Brian, "but he deserted me! He didn't ask me if it would be okay, or if I would be able to deal with it, or anything! He just went ahead and killed himself! Why would he do that?"

I sobbed so hard that my head throbbed and I could feel my eyelids swelling. It was the first time I'd cried about his death. I'd cried about plenty of other things, real tiny, insignificant things, like hearing about missing children, or watching that tribute to Sammy Davis Jr. just before he died. But I'd never realized that all those times I'd actually been crying because I missed my father. It was as if it had just happened, and it was so unbearably traumatic that I felt dizzy and sick. I was aware that Brian's arms were wrapped around me, and as if from far away I could hear him saying, "That's good. Go ahead and cry. You'll feel better."

And so I did. Grateful that he wasn't shushing me, I cried and cried and cried. I clenched my fists and swore and cried some more, until I was completely exhausted and drained. Then Brian lay on his back and pulled me tenderly into his arms. It seemed like hours passed before I pulled myself together enough to speak in a steady voice.

"I hated him for doing that, but I never realized it until now. I thought I hated my mother. But my mother is the one who stayed with me. She was the strong one, he was a coward. My poor mother! I've been so mean to her! How could I do that to her? After all she did for me? She put me through college, and supported me until I got a job. God! Do you think she'll ever forgive me?"

"I'm sure she will," Brian soothed. "Don't worry." He stroked my hair. I was so weary that I thought I was going to throw up. I felt Brian kiss the top of my head. And then I dropped off to sleep.

CHAPTER NINETEEN

When I woke up, the pink rays of dawn were streaming in the windows. Outside, I heard the birds beginning their day. In the summertime birds like to sing early in the morning while it's still cool, then they quiet down as the temperature rises; come evening, they resume their chirping. As I listened, I could hear a mockingbird giving an impressive display of different calls. To the untrained ear, the mockingbird sounds arbitrary, even frantic. But my father had taught me to discern a pattern, and for several minutes I listened with pleasure. I thought back to the apartment I used to share with Alex, and the mockingbird who lived nearby. Sometimes I'd be awake at 3:00 in the morning, and I'd hear him singing his heart out. Other times when I came home from work, I'd see him in tree and I'd stop and whistle and say, "Mock me! Mock me!" which never met with any success.

I gradually became aware of my surroundings, and reflected on the previous night. My eye lids were so swollen that my eyes stung and wanted to stay shut. But in a way, it was a good feeling. Obviously, I'd never completely come to terms with my father's suicide, and it had been chewing at my insides for all those years. That was why I'd been so chronically depressed

— because unconsciously, I was still mourning. But I felt calm and relaxed, like I'd just finished a war and didn't have to fight anymore. And there weren't words to describe how wonderful that was. I couldn't remember the last time I'd felt so peaceful as I felt at that moment, lying in Brian's arms, listening to that mockingbird.

I raised my head, and the movement woke Brian. He opened his eyes and was startled to see me there, draped across his chest. But then he smiled, slowly and blissfully. He reminded me suddenly of a picture my father had shown me once of Ralph Waldo Emerson, with a benevolent expression and eyes full of unconditional love. I thought that if Brian's family saw him at that moment, they wouldn't even recognize him.

"Good morning," he said tenderly, giving me a little squeeze.

"Good morning."

Shyly, he kissed my forehead. I recalled the promise I'd made Howard.

"I'm sorry we didn't . . . I mean, I'm sorry I fell asleep," I apologized awkwardly.

"That's okay. I guess you needed to. You dropped right off."

"Usually it takes me a couple of hours," I said.

"I watched you sleep. You're really beautiful."

I studied him. Was he going to want to have sex now? I really didn't feel like it, but of course I'd told Howard . . .

"I don't mean physically," he said, then added hastily, "I mean, I think you're attractive and everything. But your beauty goes deeper than that. You know why?"

I shook my head. The praise sounded odd coming from him. If he'd said those things yesterday, I would have laughed or been offended. But he sounded sincere.

"Because you're so full of love. You loved your father so much . . . and deep down, I think you love your mother just as much. And I think you still love Alex. And in a way . . . I feel like you love me, too. I mean, why else would you have come to me last night? Even though I'm overweight and a diagnosed psychotic, you weren't afraid to spend the night with me. You don't know what that means to me. For years, I've been trying to figure out what it is that other people have and I don't. You know what it is? It's love. I don't love anyone. I don't even love myself. But I just lay here last night, watching you sleep, and I thinking about it, and I thinking about my family, and I realized that I love them. I do! And I never even knew it before! Can you believe it? But they've always been so good to me. They never complain about supporting me, they never tell me to lose weight or get a job or anything. They help me when they can, and accept me when they have to. I never appreciated them until last night."

I was staring at him in shock. I never would have believed such a transformation could take place. It was as if he'd found the answer to a question he'd been asking all his life. And it had happened without sex, too. It had happened because for once Brian had felt needed. For once, someone had turned to him in despair, and he'd been able to help. He was feeling as fulfilled as I was. The two of us were grinning from ear to ear.

"Let's go downstairs," he said.

I agreed, and we swapped smiles. Outside, the mockingbird kept singing.

When we reached the dining room, we heard voices. I started to go right in, but Brian held me back and opened the door just a crack so that we could peek in. Howard and Hattie were sitting next to each other, holding hands. The newspaper lay next to Howard, unopened. Once again, they reminded me of newlyweds.

"Aren't they incredible? They've always been like that," Brian whispered.

"I wonder how they keep it up," I said wistfully, thinking of the abrupt end to my own parents' marriage. I thought about Alex and wondered what we would have been like as a middle-aged married couple. It bothered me to think I would never know. I thought back to the way his eyes used to follow my face and my gestures whenever I spoke, and I remembered the warm feeling it used to give me, to know that he loved me so much. I used to think he watched me the way a dog watches its owner — steadily and affectionately, and ready to do whatever it's told. A lump formed in my throat, and for a moment I felt as if I would cry again. But then I just took a deep breath and promised myself I'd think about it all later.

Brian pushed open the door and we walked in.

"Morning!" Howard greeted us cheerfully. His eyes searched mine, then Brian's, then mine again. I hoped to God he wouldn't say, "Well! How did it go last night?"

"Good morning, you two," Hattie said with a sweet, maternal smile. Suddenly I missed my own mother so much that my eyes misted and a lump in my throat ached.

"Hey, Mom and Dad," Brian astonished them by putting an arm around each of them and saying humbly, "I just wanted to let you know that . . . I've been a real shit."

"You haven't been a . . . what you said, Sweetie," Hattie stammered.

"I have so been. And I apologize."

"No need to, son," Howard said, and his hearty voice was for once feeble with wonder.

Embarrassed, but relieved and pleased, Brian patted their shoulders, then stood and let his hands slide to his sides. Howard and Hattie were staring, stupefied, at one another. I saw that Brian was perspiring a little, and trying to be casual.

"Something to eat?" he asked me.

"No thanks," I said. I was anxious to leave the three of them alone. "I think I'll go upstairs and have a nice, hot bath."

I turned and ran out. I was so happy I felt like I would pop. And as I lathered up with Trixie's fragrant bubble bath, I felt even better. I borrowed her razor and shaved my legs, and shampooed my hair, and hoped I would be able to find a Painted Bunting soon, because then I'd be able to go home and see my mother.

CHAPTER TWENTY

"You're sure you won't visit with us for a few more days?" Hattie pleaded as we headed toward my car — André had brought it around front. I gave her a grateful, regretful smile, and shook my head.

"Thank you anyway, but I have one more thing to do, and then I'm driving home."

"Will you come see us again?"

I said of course I would, but I think we all knew I wouldn't. For one thing, I didn't know if I'd ever be in the area again. Not only that, but I knew that the closeness of the night before could never be duplicated. I had the feeling that if I ever saw them again, we wouldn't have anything to talk about. We'd shared our lives for twenty-four hours, and we'd all gained something. We didn't need to be together again. I think they felt that way, too, because a peaceful silence accompanied us to the car. Brian put his hand on my back for a moment, then let it fall. André opened my door for me, and I saw, sitting on the seat, my poor, lonesome, dirtless fuchsia.

"Hattie, I have a present for you," I said impulsively.

"Oh goodie! What?" She was delighted and peered eagerly over my shoulder.

I retrieved the fuchsia and handed it to her. "It's a fuchsia. It needs more soil and some water. It's really a nice plant."

"Well thank you, my dear!" Hattie exclaimed, accepting it and examining it with pleasure. "I'll take very good care of it, and think of you whenever one of the servants waters it."

"Good enough," I said.

"Would you like Goliath in return?" she asked hopefully. But as Brian's face went white, she forced herself to giggle and said, "Just teasing, honey." Then her face lit up and she cried, "Oh, *I* know! Take this, my dear."

To my amazement, she slipped off her pinkie diamond ring, the one that would have paid my rent for a year, and handed it to me. I was so astonished that I just held out my hand and accepted it.

"I can't take this," I said.

"Why not? I want you to."

"Because I . . . I . . . ." I couldn't stop myself from putting it on, and watching it sparkle brilliantly in the sun. Where would I ever wear such a thing? "I just don't think . . ." I looked back at Hattie and saw that she was beginning to be hurt. "I don't know how to thank you," I said.

Her smile shone like the diamond. "No need to thank me," she declared. Before I could say another word, she pulled me into a tight, perfumy hug. "Goodbye, my dear. I enjoyed having you here *so* much!"

"I enjoyed it, too," I said, "thanks for everything."

She released me so that Howard could hug me. As he did, he whispered in my ear, "Thanks for the favor, Lark. Guess sex did the trick, huh?"

"Nope," I said, "Brian did this all by himself."

Obviously, he didn't know what I meant by that. I wondered if Brian would ever tell anyone what had happened between us. With a smile, I pulled away to hug Geoffrey, and finally Brian.

"Talk to your mother," he advised confidentially into my ear.

"I'm going to," I said. We kissed, then I climbed into my car and backed out.

"Goodbye! Goodbye!" they all called. As I drove off, I saw that Brian had his arm around both this parents.

"Not a bad night's work," I remarked to my fuchsia. Then I remembered I'd given it away. "Now why did I do that?" I wondered aloud. "I'm going to miss it. Oh well, it'll be better off with Hattie in that beautiful home. And she'll give it more soil."

Turning on the radio, I heard, ". . . *was the new one by Food of the Gods. Haven't heard from them in a long time! We're glad they're back together. Back in a minute with your favorite music . . .*"

"I knew it! I knew it!" I shrieked, steering unsteadily and slamming my palm on the dashboard. "I knew it had to be them!" My hand fumbled desperately with the knob, constantly turning it in search of the song. I was glad no one was in the car with me, I would have driven them nuts. For hours I drove like that.

Anxious to get to North Carolina, I didn't stop for lunch or dinner. From Route 81 I got onto 83, which took me straight through Virginia and across the border. I was so excited that I was wriggling, and I even quit playing with the radio for a little while as I tried to see everything at once. I assumed that I'd arrive in North Carolina, see a Painted Bunting, then turn around and go home. But of course things never turn out the way we plan.

For one thing, by the time I got there, it was too dark to see any birds, never mind crawl through briar patches and swampy thickets in search of a specific species. So I decided I might as well find a place to stay for the night, then spend the whole next day looking.

And then I saw it. A white brick building, three stories high, with the call letters **WCLX** painted across the front. A radio station! I stopped my car so abruptly that I almost went through the windshield. What luck! What incredible luck!

"I can ask someone there if they know anything about the new Food of the Gods album," I said, delirious with joy.

I got out of my car and slammed my door. There was no one else around. Up the road a bit I could see houses, and everyone had their lights out. In one window, the blue lights of a television set flickered. It was very peaceful and quiet. I began to walk, and a dog, alerted, barked and barked until someone let it out. I started to feel a little self-conscious about stopping at a radio station in the middle of the night to ask about the new Food of the Gods album. I mean, wasn't that just a little bit strange? Nevertheless, it was what I wanted to do at that moment. I didn't care how late it was, or if anyone thought it was weird. I knew that if I didn't stop, I'd be pissed at myself the next day. So I climbed up the steps of the radio station and reached out to open the door.

For some reason, I assumed it would open. When it didn't, I took a second to try to figure out what time it was. My best guess was that it was well after midnight, maybe even after one. I tugged on the door, as if maybe I was mistaken, maybe it wasn't locked. But it went CLAK CLAK CLAK and stubbornly refused to yield. My hand fell to my side, and I did nothing but stand and stare with my mouth open.

"Well now what?" I muttered, completely at a loss. There was a light on inside, so I cupped my hands around my face and peered in. I saw a large desk with the call letters **WCLX** stenciled on the front. I looked to the left, and to the right, but all I could see were dark hallways. Heaving a deep sigh, I turned my back to the door, leaned against it, then slowly sank dejectedly to the ground.

The next thing I knew, someone was shaking me, saying, "Miss? Miss? Are you okay?"

My eyes opened drowsily, attempted to focus, then grew wide and stared.

"Who are you?" I demanded.

"Johnnie Matthews. I'm a DJ here at WCLX. Who are you?"

I tried to collect myself. I looked around, wondered why I was so chilly, then realized that I was outside. Then I wondered what I was doing outside. I tried to rise, but a crippling pain shot through my back, and with a little cry, I sat back down. And suddenly it all came back to me, and I blinked at him rapidly.

"A DJ? You're a DJ?"

He nodded. I tried to get up again, and he reached out to help me. I was stiffer than hell, and winced with pain as I studied him. He was tall and well built, with dark features, and a very kind face. Most impressive, of course, was his voice, which had a slight, leisurely southern drawl, and was sexy as a saxophone solo. Most of the hairs on my arm stood up at the sound of it.

"You okay?" he asked.

I nodded. He nodded, too, and while he unlocked the door, I told him who I was and why I was there.

"Yeah, that new Food tune is excellent!" he enthused as he let us both in. "I play it as often as I can get away with."

"So . . . Food is back together again?"

"They seem to be. They changed labels — they're not with Zener Bros. anymore."

"Really? How about that. Wonder what happened."

"Apparently Zener dropped them. Said they weren't making enough money. But that's probably because they didn't get enough promotion."

"Probably," I agreed. We'd walked past the desk, down a hall, and found ourselves waiting at an elevator. It was kind of funny, the way he'd just let me come in and follow him like that. Meanwhile, throughout the building, we could hear music and the voice of the DJ who was on.

"*. . . almost time for the sun to come up. Highs today will be in the mid-nineties. Another scorcher, here at WCLX, where we play all classics!*"

Johnnie pushed the button for the elevator, then, when it didn't arrive immediately, pushed it twice more as if to hurry it.

"Can you play that song for me?" I bubbled. "Can you dedicate it to me?"

"Sure could! What did you say your name was?"

"Lark DePaolo. My father named me after the bird," I said.

"That so?" he said, politely feigning interest. I didn't care. I was going to hear the new Food tune. And it was going to be dedicated over the air waves to *me*!

Musical chit chat accompanied us to the third floor, through the elevator doors, and down the hall.

"*. . . Here at WCLX we play songs by your favorite classic artists. Why? 'Cause we LIKE you, that's why! Coming up next, brand new Springsteen . . .*"

We reached a room with a big window, through which we saw a young blond guy speaking into a mike. Above, a lit sign said ON THE AIR. It was very exciting. I'd never been in a radio station before.

"That's Mickey Moore," Johnnie told me. Mickey saw us and waved. We waved back, then continued down the hall to another room with a window. Johnnie let us in, flipped through some singles in a rack, located the one by Food, and said, "Here it is. I'll start my show with it."

"Great!" I said. "Is the album out yet?"

"Not yet, but we're expecting it any day now. I can't wait to get it."

"Me neither."

". . . Be leaving you for now. Johnnie Matthews is up next to play more of your favorites, right after this. You're listening to WCLX, where we play more music than anyone!"

There was a commercial and a couple minutes of news. Apparently, it was 5:00 in the morning. While we listened to an early traffic report, Johnnie pulled over a chair for me, then seated himself. I watched as he positioned the single on the turntable, and clicked on his mike.

"Morning, folks! It's about six after five, and you're listening to WCLX. Johnnie Matthews here, to play all the music of your favorite classic artists. I have a terribly charming young lady in the station with me. Lark DePaolo, named after the bird. She's got a request for a song by her favorite band, and mine, too! Here's the new one by Food of the Gods . . ."

He released the record, and the music started right up. What an amazing song it was! It had a catchy, upbeat chorus about how the days used to drag — *"Seemed the time went on and on/Phone never rang and all hope was gone"*, and then it had this bridge that was real soulful — *"Just when I thought that everything was finished/Music came along and everything was different/My whole life changed with the strike of a chord/And suddenly I knew what I was living for."* Johnnie and I smiled broadly at one another, and listened it through. When it ended he released a conspiratorial chuckle and said, "I'm breaking WCLX policy rules and my program director is going to be pissed, but hell, who's listening to the radio this early in the morning?" and played it again.

"Will you get in trouble?" I hissed, scandalized, but delighted.

"Nah — I'm Johnnie Matthews."

I said, "Oh," and laughed; not because I thought it was funny, but because I felt so good inside. I'd left my old life behind, and Food of the Gods was back together again.

We listened to the song again, and for the first time, I was aware that drummer Burley Dunmore had really outdone himself on one particular lick. And during the final chorus, Chris West on keyboards executed an impressive Chopin trill. A perfect conclusion to a song that was a tribute to music. "What an amazing band," I murmured.

Johnnie nodded, then leaned toward the mike. *"That was the new one by Food of the Gods,"* Johnnie said casually when it ended the second time. *"Album's not out yet, but should be very, very soon. Watch for it."*

He'd cued up a commercial, and the second he stopped speaking, a man's voice came on, praising his new car. Johnnie lowered the volume, then turned in his seat to face me. The phone rang abruptly, startling us both, but he didn't answer it.

"Thanks so much," I said shyly.

"My pleasure, Lark." He sounded real sexy saying my name. I thought back on Alex's sweet voice, and a little pang of anxiety shot through me. I rose, and Johnnie did, too.

"Find your way out okay? Wait — Mickey can walk you down." He motioned at the window. I turned and saw Mickey

Moore standing there, waving back. Johnnie indicated that he was to enter, which he did.

"Mickey, this is Lark. Can you walk her out?"

"Glad to," Mickey grinned amicably. "Come on."

I thanked Johnnie again, and he had just enough time to smile before going back on the air.

"We're back. It's gonna be a perfect day — perfect if you've got air conditioning! Ha ha ha! Here's an oldie but a goodie — 'Sloop John B' by the Beach Boys, here on WCLX, where we play all your favorite classics!"

Mickey escorted me down the hall, into the elevator, and out. During that interim I learned that he was twenty-five, that he'd worked at WCLX for nine months, and that his second child was due in two weeks. I was impressed by all he'd accomplished in such a short time. I was five years older, I didn't have a job anymore, and I wasn't even married yet, let alone having children. "Good luck," I said, but I knew that luck didn't have anything to do with it. He was living a good life because he was letting himself be happy. It was time I let myself be happy, too.

CHAPTER TWENTY ONE

The sun was up by the time I reached my car. Once again, I could hear the birds welcoming another morning. And it suddenly occurred to me that I didn't have any idea what day it was.

"Oh well, guess it doesn't matter," I said happily as I made my way back onto a main road. I drove for a while, keeping my eyes open. Wide open. I really didn't have the slightest idea about how I was going to find a Painted Bunting. All I knew was, the more I thought about seeing one, the more excited I became. It's such a fabulously bright, colorful bird, with a red breast, a green back, and a purple head. I can still remember the first time I saw a picture of one. It was in a book of birds that my father had, and he pulled me into his lap, opened up the book, and even before he pointed to the bunting I said, "Ooh, Daddy, what kind of bird is that?" He told me the name, and we made an immediate agreement that we'd go find a real one some day. And as I drove along, I thought about how nice it would have been to have him there with me. But he was gone. I was going to see a Painted Bunting, and he wasn't, and it was his own damn fault. Suddenly I wished I had a camera so that I could take a picture and show it to my mother.

I travelled down some smaller roads, and found a moderately wooded area. Pulling over, I stopped the car and got out. The air was magnificently fresh, and not too humid yet. I took a deep breath of it, then another, and realized all at once that I felt great. I was humming the new Food tune, and my heart was thumping with anticipation. I couldn't recall the last time I'd taken a walk in the woods.

"I need to do this more," I announced. "On weekends, or in the evening after I've gotten home from work. Instead of watching TV, I can go out. And hey, I bet Alex would . . ."

My voice died out, and I stopped moving. I bet Alex would *what*? Forgive me for walking out on him without a word? My father forgotten for the moment, I resumed walking, slowly, and tried to come to some kind of decision. It was time to figure out what I was going to do next. How was I going to live without Alex? Would I have to? Or would he take me back if I asked him to? Did I want him to? I needed to decide if I still loved him. I knew that when I left, I didn't feel like I did. But at that moment, I missed him, and realized that I wanted to go home to him. But would he understand? What if I told him, "I left because I was fed up with my life, and that included you, but now everything is fine, and I want us to get back together." What would he say? Would he tell me everything was fine, that he wasn't at all upset? I doubted it. He had to be pissed. He *had* to be. Who would I marry, if not Alex? Bradley Crenshaw III? Have a son named Bradley Crenshaw IV? It didn't sound too appealing.

The ground was getting damp and soft. The air smelled very thick, and I had to breath through my mouth. A light film of sweat made my sweatshirt and shorts cling to my skin. My pink Nikes were spattered with wet dirt. My fuchsia would have loved this soil, I reflected, suddenly feeling like the loneliest person in the world. I walked some more. All the while, I was aware of all kinds of wildlife. I saw a variety of unremarkable warblers, a bright Summer Tanager, and heard, but didn't see, a White-eyed Vireo. Permeating all the other calls, I could hear Carolina Wrens. My father once told me that Carolina Wrens often provide the backdrop noise in movies. Their calls are dubbed into a soundtrack whenever the director wants to create a meadow or field kind of atmosphere. It's funny, because sometimes it's not particularly appropriate, like if the movie takes place in California. My father and I used to try to distinguish if the calls were genuine or dubbed in.

I was really enjoying the sensation of the twigs snapping beneath my feet, and the occasional bright ray of sun streaming in through the trees. I couldn't hear any sounds other than the birds, the frogs, the crickets, and my own footsteps. It was incredibly peaceful. And all of a sudden I felt so in tune with nature that I thought I would cry. It seemed that, in the face of all those glorious trees and that rich, woodsy smell, all my problems seemed pretty trivial.

Now I'd like to say that at that moment the elusive Painted Bunting fluttered by. But of course it didn't. I kept walking. My feeling of oneness faded a little, because it was too intense to last for very long, and suddenly my stomach rumbled. I tried to recall when I'd eaten last, and couldn't. Reluctant to leave before I'd seen the bunting, but even more reluctant to starve to death in the middle of some North Carolina woods, I walked back to my car and drove out, looking for a restaurant.

There was a small run-down diner built to look like a boxcar along one of the secondary routes, so I pulled into it.

There were a lot of people inside. I realized it had to be about noon. Seating myself on a tall swivel stool with a red cushion, I admired some Danish under glass, then nodded at the waitress who greeted me.

"Hot enough for you?" she asked. She kept pushing a strand of her straight, fine, light brown hair behind one ear. Her bangs needed to be trimmed. But her eyes were very warm. All in all there was something very friendly and comfortable about her. I said it was, and asked if it was hot enough for *her*. She laughed as if I'd said something terrifically clever, and I laughed with her for no particular reason.

"What'll you have today."

"Well, what looks good?"

She selected a menu that was tucked into a menu holder, and flipped it to me. The metal tipped pages went CLIK when it hit the counter. I thanked her, and opened it up to have a look at my options. I was aware that she was waiting, not impatiently, with her hand hovering over her pad. I saw that her badge said, "Hello, I'm MAGGIE" and took a moment to consider the repercussions if everyone in the world was required by the law to wear a name badge. I wondered if people would be friendlier if they knew everyone else's name, and if there would be as many murders.

"Let's see . . . definitely one of those Danish . . . and something cold. Maybe a fruit cup and some cottage cheese. And ice tea with plenty of ice."

"Mm Mmm!" Maggie enthused as she wrote everything down, then left to execute the order. I sipped the sweaty glass of water she'd set before me, and pivoted slightly in my seat to look around at the other patrons. Next to me was the thinnest man I'd ever seen, thinner than Gandhi even, dragging on a cigarette and reading a local paper. The headline said, *Girl Raped By Drug Dealers*. I took another sip of water and tried to imagine what it would be like to be raped. By drug dealers.

Unwillingly, I thought back about the last time Alex and I had made love. One thing you could say about Alex — apart from falling asleep right after sex (I'm told it's a biological phenomenon, and that most men suffer from it) he was considerate in bed. He always made sure I was enjoying myself. Once I was dating this guy who was so hung up on his looks that I wasn't even attracted to him, he was just *too* handsome, and whenever we were in bed I tried to make it obvious that I wasn't interested in sex with him. I sighed a lot, and didn't initiate any action. But he never once even suspected.

Next to the Danish under glass was a display of postcards. *I ♥ North Carolina. North Carolina is for ♥ers. I left my ♥ in North Carolina.* I thought they'd pushed the ♥ motif a bit too far. Nevertheless, I took one and absently turned it over. *Love and sun from the land of fun!* I thought about sending Alex a postcard, explaining why I was in the land of fun without him.

With a sudden SNAP! the guy next to me adjusted his newspaper, and a new headline was revealed: *Crime Rate Soars.* I took another sip of my water and thought about that. Back home, in Boston, we were approaching a record number of murders for the year. It was alarming. I didn't live right in the city, so it was just something I kept hearing about, but hadn't experienced for myself. I thought about the guy who'd stolen my car. I wondered if he'd have killed me if he'd gotten the chance.

"Here y'are," Maggie's cheery voice broke into my musings. I looked up and saw her standing there, grinning and holding my cottage cheese in one hand and my fruit cup in the other. She seemed delighted to be serving me.

"Are you always this nice?" I asked her.

She chuckled, shrugged, and nodded.

"Sure. Why shouldn't I be? Got a job I like, a wonderful husband, and two great kids. Here, look." She set down my plate of cottage cheese and my fruit cup, wiped her hands on her apron, then reached into the back pocket of her uniform and withdrew a couple of snapshots. "I always carry these with me. Anytime I think about how I'm working my butt off and not getting paid much, and how my husband's been out of work for six months, and my kids don't have any nice clothes, or when a customer is rude or something, I just take these out and look at them, and I feel better."

She held them out to me. I took them and saw that the first was a picture of two children, a boy and a girl. They both wore Maggie's bright smile, and they were even holding hands. The boy had on shorts. His knees were round and soft, like that biscuit dough that comes out of a roll. The girl was in a dress which looked itchy and uncomfortable. But she didn't seem to mind.

"They're adorable," I said. I'd never been the type of person who liked to look at pictures of other people's kids, but Maggie's were different. I looked at the other snapshot, and saw that it was one of those family portraits you get coupons for at Kmart. Maggie's husband was so chubby that the buttons of his shirt across his belly were just about to pop open. But there was something friendly about him, and I couldn't help smiling. Maggie was dressed up in a jade green blazer and matching slacks. The little girl was wearing the same dress she'd worn in the other picture, but the boy was dressed in a suit. He looked unnatural, almost like a mannequin. "Those are really nice, Maggie," I said sincerely, handing them back to her. She took them, glanced fondly at them, then returned them to her pocket.

"Thanks. You got any kids?"

I shook my head. Alex used to talk about having kids, but since I'd never committed myself to marrying him, making a decision about kids had always seemed kind of unnecessary. I wondered what kind of mother I'd be. Probably terrible. I used to let my dog Elliot get away with anything. And he'd been about fifty pounds overweight because I couldn't resist giving him a portion of whatever I ate, including ice cream. He always got to lick the carton when I was done with it.

"Oh, you should *have* some!" Maggie urged. "They're *so* much fun! I don't know *what* I'd do without Jessie and Millie!" She had a very lively manner of communicating, using lots of hand gestures and emphasizing a lot of key words. I enjoyed watching her and listening to her.

"I don't have anyone to have kids with," I told her. With my fork I kind of pushed the curds of the cottage cheese around in the dish. For some reason, I felt like it would be rude to eat in front of her, even though I knew that people ate in front of her all day long.

"Not even a boyfriend?" she demanded, amazed. "Wow. How come? You're so pretty! God! I'd *love* to look like you!"

I was caught off guard by her praise. I'd been so fascinated by her sweetness that I hadn't considered her physical appearance at all. But then I noticed that she was pretty heavy, and although she was probably about my age, her brown hair had more grey in it than mine did. She probably thought I was real slim and sophisticated. I'd always been told I was "pretty." But I wondered if people thought I was nice. I thought about my funeral again, and tried to imagine how people would describe me after I'd died.

"How can you say that. I look terrible," I protested,

indicating that my hair was a mess and that I wasn't wearing any makeup.

"Because you look mysterious," she said.

"I do?" I was pleased. I must really look different now, I thought, now that I've spent some time on the road. I pondered this for a moment, then Maggie reached over, lifted up the glass, selected a Danish for me, and set it on my plate of cottage cheese, carefully, so that they wouldn't touch each other.

"You eat your cottage cheese and your fruit cup and your Danish. I'll be right back with your tea."

"Are you going to wait on me or *what*," a voice nearby said irritably. Maggie and I glanced over and saw a guy seated a couple of stools away, frowning and glaring.

"Be right there, sir!" Maggie called out gaily. She sent me a glance and rolled her eyes. I couldn't keep from giggling.

As she went over to him to take his order, I fell to eating my cottage cheese and my fruit cup. A moment later she returned with my tea, but another customer was calling her and she didn't have a chance to chat with me again.

After I'd eaten everything, including that fattening Danish, I reached into my purse for my wallet. A glimmer caught my eye, and I realized I was still wearing the diamond pinkie ring Hattie had given me.

"All set here?"

I looked up, and there was Maggie, smiling and preparing to clear away my dishes.

I said, "Maggie, I want to give you something a very dear friend gave me. Would that be alright?"

She was surprised, and stopped working to regard me curiously.

"Really? What."

"This." I slipped off the ring and handed it to her. Her

eyes grew wide. Under the overhead lights, it sparkled like water when the sun hits it just right.

"God! I can't take this! This must be worth a *fortune!*" But she put it on and kept staring at it, admiring it and turning her hand this way and that, catching the rich glints.

"I want you to have it," I said, "in case you need it."

"Aw, honey, I didn't mean to sound like I was begging or nothing. I mean, my family and me, we'll get by." She started to take it off. But I put up my hand to stop her.

"You didn't," I assured her. "But I don't have enough money for a tip. So I have to give you that, instead. Okay?"

"Well . . ." she was still doubtful. But like a child she was fascinated by the ring. She'd probably never seen such a big diamond in her life. Actually, I hadn't either. But it didn't seem to mean as much to me. Finally she made up her mind. "I don't know why you'd want to give away something so beautiful, but, well, thanks!"

She reached across the counter and gave me a vigorous hug. It seemed kind of funny. I was aware that the other patrons were staring at me, and a week ago I would have been very self-conscious about it all. But I hugged her back and said, "I'm glad you like it."

"I'm going to wear it every day," she promised, releasing me. I settled back into my seat and finished counting out the money for my meal. Someone called for her, but she ignored him. I could tell she was staring at me, trying to figure me out. "I still say you're crazy to give it away."

"Not at all. I'm just grateful for a friendly face. You know something, Maggie? I think the newspapers only have half the story. I think there are a lot more nice people out there than they're telling us. I think maybe it's time we started emphasizing what's *good* about this country instead of what stinks."

She was surprised. I guess she hadn't expected a discourse on the state of the union. But I got to thinking about all the people who had helped me find myself and determine what had gone wrong with my life. I hadn't done it by myself. There'd been strangers who'd pushed me in the right direction. And there was no way I could go back and thank them all. All I could do was continue the trend.

"Thanks for the meal, Maggie," I said, rising.

"Thank *you*," she said. I could see she was still puzzled. I looked around and saw that people were still listening and staring. I must have seemed pretty odd to them, coming into a run-down diner built to look like a boxcar, and leaving a diamond ring for a tip. But Maggie's family was in for one hell of a story when she got home that night.

CHAPTER TWENTY TWO

As I headed out of the diner, I was suddenly so sluggish I could barely walk. For a minute I didn't know why. Then I remembered that I'd spent the night before sleeping on the front steps of a radio station.

"Think I'll get myself a nice room, grab a nap, then go out again early in the evening and look for the bunting," I said merrily.

"*. . . Situated with easy access to three major highways, Highland Estates Condominiums are both attractive and affordable . . .*"

"And if I don't see one, I'll get up real early tomorrow morning and look some more," I went on.

"*. . . My hair used to be limp until I discovered WONDER shampoo. Now Bob loves the way it feels!*"

"And if I still don't see one . . . then what? I can't stay here in North Carolina forever, can I? Or can I? Call my mother and tell her I love her . . . then get a job . . ."

"*. . . Coming up next, the new one from . . .*"

"Yes? Yes?" I turned the volume way up.

"*Fleetwood Mac.*"

I turned the volume back down. A moment later, I spotted

a motel and pulled right in. I located the office, went in, and saw a red-headed woman about my age, irreverently snapping gum and ploughing through a Best Seller that had been made into a really terrible movie. Automatically, my eyes went to the sign on the desk, which said, "Pebbles O'Brien, Manager."

"Hello," she greeted me, setting the book face down and probably destroying the spine. "Help you?"

"Is your name really Pebbles?" I heard myself ask. I saw that her skin was very fair and very smooth, like a picture on a skin care product. Her eyes were bright green, and reminded me of Alex's. She laughed at my question, and it had a pleasant, musical quality. I couldn't help smiling.

"No. Molly. I've been called Pebbles for years because of my red hair. Did you ever used to watch the Flintstones?" When I nodded, she went on, "Remember their daughter, Pebbles?"

"She had red hair," I allowed, "but she wore it up. In a bone."

"Well now, so she did. I'd forgotten that." She laughed again. Then she asked if I'd be needing a room. I noticed that she had the slightest hint of an Irish accent, and I was envious. Accents are so exotic. My Boston dialect was nothing to brag about.

"Yes."

"Okay. Let's see, then . . . we have lots of vacancies. What's your favorite number?"

"I'm not sure."

"How about six?" she smiled brightly and handed me the appropriate key.

"Six sounds good, I guess," I said, taking it. We studied one another. I wondered what she was thinking about me. I wondered if she admired my dark hair and tan, since she was so pale — sometimes that's the case. I thought her freckles were adorable, but I wondered if she wished she had my clear, uniform complexion. I don't know how long we stood looking at each other, but we were both smiling, and we liked each other at once.

"Lark DePaolo," I said, extending my hand. She nodded and shook it.

"By yourself, are you?"

I said, "Yeah," and to my surprise, explained that I'd walked out on my boyfriend because our relationship had kind of gone stale.

"I know exactly what you mean," she said. "I think I'm going through that right now, too. My boyfriend, Chris, and I have been dating for two years. He doesn't buy me flowers anymore, or send me cards. Not that he doesn't still love me . . . but I guess the romance has kind of died. Why does that always happen? Why do men always forget to do little things like that? Why aren't they ever as sweet as when you first started to date them?"

I shrugged, but I was thinking guiltily of all the nice things Alex used to do for me. He'd pick up little things for me at his drugstore, like candy bars or ribbons for my hair, and a couple of times he'd left work early without telling me, to sneak home and prepare a nice dinner, complete with wine and candles.

"I don't know," she sighed. "Sometimes I'm not sure if I'm still in love with him. Do you know what I mean?"

"Uh huh."

"I mean, I know I love him . . . but sometimes I feel like I'm not *in* love with him. He's asked me a couple of times to marry him, but I keep telling him I have to think about it."

"Have you ever told him how you feel?"

"Are you kidding?" She laughed, and I had to smile, even

166

though I didn't see the humor. "Can you ever tell men *any*thing?"

"Probably just a phase you're going through," I said, recalling all the times I'd been upset about real dumb stuff, and Alex had listened to me whine, and had apologized without getting all defensive. It was one of the things I loved best about him. "I bet that if you broke up with him, you'd really regret it." I should know, I thought.

"You think so? I don't know. Sometimes it's so hard to know what you want."

"Boy, that's sure the truth," I agreed. "Probably the only way to find out if you want something is to deprive yourself of it for a while and see how you feel."

"So you think I should stop seeing him for a while?" Her eyes scanned mine for an answer. I was alarmed that I'd inadvertently given her advice. What would Chris think, if he found me telling his magnificent Irish girlfriend to break up with him?

"No, no, I didn't say that," I told her hastily. "I don't even know the guy! It would be impossible for me to say if . . ."

"Hey! I have an idea!" she interrupted. "Maybe . . . no, that's asking too much."

I couldn't resist her beautiful green eyes, staring into mine so hopefully.

"What," I said.

"Well . . . Chris and I are going out this evening. Maybe you could come with us, and watch us together, and tell me what you think."

I was so surprised that I fell into a chair and stared at her. Who was I now, Ann Landers?

"Sure," I said. "When."

"Eight-ish. We're going out to dinner, then we'll probably

go to a pub and hear a band or something."

It had been so long since I'd done anything like that! I nodded eagerly.

"I'd love to, Pebbles! I need to sleep for a few hours, but then I'll . . . oh wait . . . I can't." My face fell.

"What's wrong?"

"These are the only clothes I have. I've slept in them and I've crawled around the woods in them. I can't go out with you looking like *this*!" I said miserably.

"Well for heaven's sake, Lark, you can borrow something from me. We're about the same size. Bet I've got a whole closet full of clothes that'll fit you."

"Really?"

"'Course! You'd be doing the same for me, now, wouldn't you?"

I nodded. I would.

"Well, then, there you go. My room is above the office. Here's the key. Go up and find something. Then go back to your room and nap. Be back here between 7:30 and 8:00. Chris is picking me up."

"What about . . . ?" With my hand, I gestured the desk, the phone, and the door.

"My father takes care of customers when I go out. He's a real sweetheart. I don't know what I'd do without him."

"I hope you never find out," I said.

I took the key from her, followed her directions to her room and let myself in. I saw that it was very small and very feminine, with a flowered bedspread and matching curtains. She had a big colorful poster of a unicorn with long, flowing mane. I thought of the room I used to share with Alex. He'd allowed me to cover the walls with posters I'd gotten from the Audubon Society. I tried to picture any other man in the world letting me do that, and couldn't. I tried to imagine Bradley Crenshaw III saying, "Of course you can put up your Birds of Prey poster! I like looking at it, too," and had to laugh.

Somewhat self-consciously, I opened up her closet, and selected a powder blue oversized blouse, a wide bright pink belt, and a denim mini skirt that was almost identical to the one I always wore. Then I grabbed a pair of hiraccis which were my size. When I stopped back at the office to return her key and get her approval, she said my choice was fine, and that she'd see me in a few hours. Then I went to my room. But the prospect of wearing something different was so exciting that I could hardly sleep.

CHAPTER TWENTY THREE

While I napped, I had a bizarre dream. My father and mother and I were on a game show, and they'd each given an answer to some question. I thought I was supposed to pick which one was right, but the game show host told me those weren't the rules, and I wound up winning all kinds of money. It was the first time I'd dreamed of my father since he died.

I took a long, hot shower, fluffed up my hair with my fingers, and climbed into Pebbles' clothes. Sure enough, they fit me perfectly. There was a clock in the room, and I saw it was only 7:00. That meant I still had some time to take a quick walk before meeting Pebbles and Chris. Maybe I'd spot a Painted Bunting and could go home.

The area surrounding the motel was nice and quiet. I walked along a more or less main road, keeping my eyes peeled. The air was filled with bird calls. Like people, birds in different parts of the country have slightly different accents. I'm not good enough to differentiate between them, but they say a true ornithologist can identify where he is based solely on the calls of the birds. My father always claimed that he'd be able to, but of course he was gone now, and so there was no way of knowing if that was true or not.

I walked for about half an hour without seeing any Painted Buntings. Then Pebbles' hiraccis began to hurt my feet, and I hurried back, hoping I hadn't gotten them dirty.

When I reached the office, there was a middle-aged man with a very round, rosy face surrounded by a thick ring of strawberry blond hair, sitting behind the desk, reading the Best Seller that Pebbles had left behind. He closed it hastily when he saw me, and rose, wearing Pebbles' warm, welcoming grin.

"I'm supposed to meet Pebbles here," I said.

"You'd be Lark, then," he said, and his brogue was very thick. I couldn't help smiling at him as I acknowledged my identity. "She'll be down in a minute. Don't be standing there. Take a seat. Will you be wanting a cold drink?"

"Ice water would be great, if you've got some."

Nodding, he motioned that I was to sit, which I did, then disappeared through a door. He emerged a moment later with a paper cup of water. I saw there was a lime in it.

"Thanks so much," I said. "It's really hot today."

"That it is," he agreed amicably.

Just then Pebbles burst in, looking fabulous in a sleeveless sun dress with a long, very full skirt. I couldn't take my eyes off her hair. We both said, "Wow, you look nice," at the exact same time, and her father laughed.

"You're a pair, you are," he said.

"Did you meet this marvelously handsome man?" Pebbles giggled, wrapping her arms around her father.

"Go on with you," he smirked with a flip of one hand. "What time will your young man be coming for you?"

"Any minute now." Pebbles went over to the window and looked out. I was admiring her profile when suddenly a pink glow spread across her face. She looked just like a ripe peach.

171

I followed her gaze, and saw a dark haired man climbing out of a car. In the moment before he reached the office and let himself in, I saw that he had dark sparkly eyes, and was already grinning. He was very appealing in an unaffected kind of way, like a mutt not fit to appear in any dog shows, but eligible for a lot of affection, and maybe even a spot on a television commercial. As soon as he entered, his eyes sought Pebbles', and they swapped lovers' smiles.

"Chris, this is Lark," Pebbles said.

"Glad to meet you," Chris said, reaching out to shake my hand. "And nice to see you, sir," he said to Pebbles' father.

"Young man, how many times have I told you to call me Joe?" Pebbles' father growled.

"Joe," Chris obliged nervously. His eyes went back to Pebbles' face, but she was smiling affectionately at her father.

"Thanks again, Daddy. We won't be late."

"Don't you be cutting your fun short and hurrying back to me," he told her comfortably, accepting her kiss on his rosy cheek with great pleasure. I used to kiss my father like that, I thought wistfully. For a moment, I thought I was going to cry. So I pulled my eyes off them, and looked at Chris. He was gazing at Pebbles with such blatant adoration that I found it difficult to believe they'd been going out for so long. I felt another pang, recalling the way Alex used to look at me.

"We'll be going, then," Pebbles announced. With her father around, her accent was for some reason much more pronounced.

"Have fun, children," Joe said. We assured him we would, and left the motel. Chris recommended a Chinese restaurant, so we went there and he and Pebbles drank alcohol from mugs that looked like figures from a totem pole while I sipped brutal, bitter Chinese tea. The food was delicious and plentiful, and, if our fortune cookies were to be believed, our futures were bright, full of romance and business opportunities.

All the while, I was listening to Pebbles and Chris. They didn't mean to exclude me from their conversation, but there are certain things that people in love have to talk about as soon as they get together, like how their day went. Chris, I gathered, worked for a small computer company that was on the verge of a major business deal which would make them all kinds of money. However, with success comes increased responsibility and pressure. He complained about some of his coworkers, and Pebbles knew them all by name, even asked about some of them. Then Chris wanted to know what Pebbles had done that day, how many people had been by for rooms, and so on. Pebbles told him that someone named Arnold was back, and Chris burst out laughing. Apparently they shared a private joke about him. They attempted halfheartedly to explain it to me, but they were giggling so much that I couldn't make out what they were saying. So I just said, "Isn't that funny," and left them to enjoy it by themselves. It made me think of a certain customer that Alex used to tell me about, who would come into the store and tell some of the young cashiers that he'd really like to paint them in the nude.

"What?" I suddenly realized I'd been addressed.

"I said, So what are you doing all this way from home," Chris asked. Pebbles had explained to him that I was on a trip, but nothing more. I debated whether or not to go into detail about having walked out on the most wonderful man in the world, and glanced at Pebbles as if to say, "Is Chris ready to hear all that?"

"I'm looking for a bird," I said.

"What bird?" Chris had started to reach for his drink to take a sip, but paused and regarded me curiously.

"The Painted Bunting," I said. "The most colorful bird in

North America."

Chris didn't say anything for several seconds. He forgot that he'd intended to take a sip of his drink, and sat back in his seat, still staring at me.

"Son of a bitch," he said, "I know what that is. I saw one once."

"You did?" Pebbles asked, surprised. He nodded.

"Yeah. My brother and I were out in the woods, hunting. All of a sudden this bright flash went by, and when it landed, we saw that it was a bird. We couldn't believe it! We thought someone had painted it or something, it was so bright. My brother said, 'We have to bring that home and show someone!' and raised his gun."

"Oh no!" Pebbles and I said together in girlish horror.

"Well that's what I said," Chris chuckled. "So I shoved him, and when he shot, he missed by a mile, and the bird flew away. It was a couple of years later that I learned it was a Painted Bunting. I was at a party and someone had a book about birds, and I started looking through it, and there it was."

"Wow," I said, amazed.

"And you saved its life," Pebbles said, gazing at Chris with new approval and pride in her beautiful green eyes. He shrugged as if it was nothing. But it was something. Pebbles and I swapped significant glances. I shook my head as if to say, "You've got to hang onto this guy," and she nodded slightly, as if I'd spoken aloud.

"So," Chris broke into our telepathy, "what would you like to do now? There's a place Pebbles and I go to a lot. They usually have a band there."

"Sounds great," I said, still recovering from his story.

The check came and Chris paid. He was more or less obliged to pay for Pebbles' meal, of course, but certainly not mine, and I was touched. I thanked him, and he just flipped his hand, then slid his arm around Pebbles' waist. I followed them back out to the car. I was aware that they were chatting in the front seat, but I wasn't listening at all. Before I knew it, we were pulling into a place featuring a band called "Chicken Slacks."

"Hey, Chicken Slacks is a local band," I said. "I mean, they're from Boston."

"Are they good?"

"They're great." They were one of my favorites, primarily because they'd professed to be big fans of Food of the Gods. I considered relating all of this to Pebbles and Chris, then decided they probably wouldn't be interested, and climbed out ahead of them. Even before we went in, we could hear the music thumping, and it looked like everyone was on their feet, dancing.

"Great band," I said again as Pebbles and Chris joined me at the door. We went inside, and it was too loud to speak. I tugged on Pebbles' arm and motioned that I wanted to go to the ladies room. She nodded to indicate that she did, too, and we left Chris looking forlornly after us. Men always want to know why women can never go to the bathroom by themselves, why they always have to travel in a pack. They always want to know if we're talking about them. As if we had nothing better to do!

"What do you think?" Pebbles demanded as soon as we'd taken our place in line, glanced at our reflections in the mirror, and scanned the other occupants.

"I love him desperately," I said.

"I know," she said. "You were pretty impressed by his story about the Painted Bunting, weren't you." She sounded smug. I didn't blame her. Women love when other women love their men.

"I sure was," I said. "I hope you're not still thinking about needing to take some time away from him, are you?" I wasn't surprised to see her shake her head. But before she could answer, I went on earnestly, "Pebbles, I walked out on my boyfriend, and I can't tell you how much I regret it now. I mean, I *really* screwed up. He loved me and would have done anything for me, and I left him. And now I realize that I'll never find anyone as sweet and generous and considerate and easy going . . . shit!" To my horror, I started to cry, right there in the ladies room in some nightclub in the middle of North Carolina. Some of the other women regarded me with sympathetic interest, while others merely paused, then resumed re-applying their makeup. Pebbles put her arm around me.

"Don't you be wasting time crying, now," she soothed. "Why don't you just go back to him?"

"Go back to him?" I repeated, staring into her green eyes urgently. "You think I should?"

"If he loves you as much as you say, he'll take you back. I'd not worry about it."

"Really? Really?" She seemed so certain, I couldn't help believing she knew best.

"'Course. Hey . . . want to know something?" Still with her arm around me, she leaned close and said, "Next time Chris asks me to marry him, I'm going to say Yes!"

"That's great! You can marry Chris and I can marry Alex!"

We broke into giggles and hugged each other. Meanwhile, other women continued to file past us, watching us and unabashedly listening in. We used the facilities quickly, then hurried back to find Chris.

"Honey, would you mind driving Lark back to her car? She's going home," Pebbles shouted into Chris' ear.

"What?"

"Lark wants to leave!"

"What? Why?" His eyes searched mine questioningly. I read his lips as he said, "You feel okay?"

I nodded to let him know I felt fine. I felt wonderful!

"Let's go!" Pebbles cried, tugging on his sleeve. He nodded promptly, and escorted us out. As soon as the door closed behind us, we could talk at a normal volume. On the way back to the motel, Pebbles explained.

"But you won't get to see your bunting," Chris reminded me. I just shrugged.

"I have the rest of my life to see one," I said. I'm not like my father, I added to myself, I won't die without seeing one.

CHAPTER TWENTY FOUR

As soon as we got back to the motel, I dashed into my room, tore off Pebbles' clothes, and threw on my navy blue sweatshirt, my grey running shorts, and my pink Nikes with the pink shoelaces Alex had bought me.

"Thanks for everything," I said once I'd returned to the office. I handed Pebbles her clothes rolled up in a sloppy ball. With a patient smile, she smoothed them and folded them properly, and thanked me, too.

"Nice to meet you," Chris said. He reached out to shake my hand, but then changed his mind, and hugged me instead. Pebbles' father hugged me, too, then offered to walk me to the car. All four of us went out, with Pebbles and me linking arms like sisters. When we reached my Grand Am, she released me and slipped her hand into Chris'. I was so pleased for her, and so excited about going home, that I started to sniffle.

"Don't be starting with that," Pebbles' father said affectionately as he opened my door for me. "Drive carefully."

"Okay, I will. Goodbye! Thanks!"

"Goodbye!" they called.

With a final wave, I pulled out and started for home. *Home!*

On the drive back I heard the new Food tune about a dozen times. I was finally starting to learn the lyrics, and each time I sang along at the top of my voice.

"Great song!" I kept saying. "Can't wait to get the album! Wonder if it's out yet? I have to check the record stores. That'll be the first thing I do. No, wait . . . I'll go see my mother first, and apologize. Then I'll try and get Alex back. Then I'll go out and look for the new Food album."

I drove until it grew light out, and stopped briefly at a Burger King, where I had some eggs that didn't settle well. I was feeling tense inside, like I'd swallowed someone's clenched fist. I sat there, sipping coffee and looking at the telephone. Alex would probably be home. He'd be in bed, sound asleep. I thought about calling him and asking if he was mad at me, or if he wanted me back. I thought that my drive home would be a lot more pleasant if I could hear him say, "Yes! Get here as soon as you can!" On the other hand, if he didn't want me back, maybe it would be easier to hear it over the phone, instead of in person. But wait . . . maybe if I was actually there, staring with desperately unhappy eyes, he'd change his mind and tell me to stay. I went back and forth with it, trying to decide if I should call him or not, and then all at once I leapt from my seat and ran to the phone. With trembling fingers, I dug through my wallet until I located my MCI card. I had to dial about a zillion numbers before I finally got through, and then, to my dismay, I heard the familiar WHIRR, then my own voice saying cheerfully, "*Hi! Alex and I aren't here right now, but if you'll leave your name and number, one of us will call you back as soon as we can. Thanks!*" I waited until I heard the BEEEEP, then I hung up. My heart, which had been thudding, seemed to slow down too much. I had trouble breathing.

"Are you alright, Miss?" a tiny old woman behind me said.

I turned and met her eyes, which were curious and sympathetic.
"No one's home," I said miserably. She said, "Oh," and
gave me an encouraging smile. She even reached out and patted
my arm a little. I tried to smile back, but I just couldn't. In
fact, the look on her face made me feel so sorry for myself that
I just turned and sank dejectedly back into my seat. Where
could Alex be? What if he'd gone out with someone else, and
had spent the night with her? What if he wasn't upset that I'd
left him? What if he was glad? What if he'd been wanting to
ask someone else out for a long time, and now he was getting
his chance, and he was really glad? "I can't believe him, what
a *dirtbag*!" I shouted. I didn't even throw out my trash, I just
left it there on the table, and raced back to my car.

Tears were streaming down my cheeks. "He probably isn't
out with someone else, he's probably just too upset to answer
the phone," I told myself. "He probably hasn't even been able
to drag himself out of bed since I left." But of course I didn't
know for sure. And I got myself so worked up that I felt
nauseous and had to pull over. I thought that if I got out of the
car and took some deep breaths, I'd feel better, but I didn't.
Instead, I threw up all the eggs I'd eaten, and the coffee, too.

With my stomach still tense and sore, but no longer sour, I
got back into the car, and kept driving. It was late at night by
the time I reached my mother's house. I hoped she would still
be awake, but just in case she wasn't, I swung my door closed
with a loud SLAM. A light went on in my mother's room, and
I saw the curtains at her window open. She peeked out. I
waved. And then she disappeared from view. I headed toward
the front door, and she flew out to meet me.

"You're home!" she cried, holding me tighter than I'd ever
been held in my life. She hugged me so hard I couldn't
breathe, and had to kind of push her away.

"I'm home," I concurred, childishly wiping my suddenly runny nose on her bathrobe.

"Where have you been? I was so worried!"

"I'm so sorry. I didn't even think about you worrying. I was only thinking of myself. But I was so crazed that I just started to drive, and I drove and drove, and wound up in North Carolina."

She studied me. In a gesture very typical of her, she brushed a strand of hair out of my eyes. "I suppose you went looking for a Painted Bunting," she said.

My jaw dropped open. "How did you know that?"

"Because your father always spoke of going down there to see one. So was it as beautiful as you always thought it would be?"

"I didn't see one. But I saw other things. And I learned a lot about myself."

"Like what."

Comfortably, we put our arms around each other as we walked into the house. I couldn't remember the last time I'd walked with my arm around my mother and hers around me. Maybe never.

"I learned about my feelings toward you. I discovered how much I love you, and how grateful I am to you for all you've done. And how sorry I am for treating you so badly."

That stopped her dead in her tracks. I guess it was just about the last thing she expected to hear. I was pained by the astonishment on her face. Is my love such a surprise to her? I wondered guiltily.

"Really?"

"Really." For a moment, I pressed my face into her neck. She reached up with her hand and patted my face, silently digesting my words.

"I always thought you blamed me for your father's suicide," she said eventually.

"I did. But now I see that it was *his* decision. It had nothing to do with you. Mother," we'd reached the door, but before we went in, I released her to face her earnestly. "I felt horrible when I realized how much I resented you all these years. God! Can you ever forgive me?"

"Darling! There's nothing to forgive! I understood what you were going through," she said generously. We went inside and she shut the door behind me.

"But what about what *you* were going through? That must have been unbearable, to have your husband kill himself, and have your only child blame you. How did you stand it?"

"I don't know," she told me honestly, "but somehow I did."

There was some lemonade left over, and without asking, she poured me a glass. I sipped it, and it tasted great.

"Um, have you heard from Alex?" I asked, sheepishly attempting to be casual.

"He's been over every single night. In fact, last night he stayed over," she said, preparing a cup of tea for herself.

"Alex . . . Alex was *here* last night?" I demanded breathlessly.

She nodded. "Yes. There was a story on the news about a woman who'd been missing a week, and they'd found her, murdered. I was really upset, and he stayed here so that I wouldn't have to be alone. And then he came back after work. I made him dinner. He left about an hour ago. I don't know what I would have done without him! He knew how worried I was, and he was so sweet and attentive. We spent the whole week talking about different things . . . about you, about your father, about his parents. A couple of times we cried together, because we were so . . . "

"You and Alex *cried* together?" I interrupted, amazed. I'd never seen Alex cry.

"Well for heaven's sake! He's been worried to death about you! He assumed he'd done something to make you walk out. Know who he reminded me of?"

"Daddy?"

"No. Me. He was blaming himself for your disappearance. When your father committed suicide, I thought it was because I hadn't been a good enough wife. But it wasn't that at all. It was because he was irresponsible and unpredictable. *You're* the one who reminds me of Daddy."

"Oh." That was criticism, but it was justified, so I had to accept it. "Mother?"

"Yes?"

"I mean . . . you loved Daddy, didn't you?"

"Very much. Too much. That's why I hated to see the two of you go off like that. Not because I didn't like to see you have fun together, but because I was afraid you'd never come back."

"Well of course we would have come back!"

"Yes, but I didn't know that."

"But . . . would you really have divorced him?"

My mother didn't answer right away. She sipped her tea, then examined a nail that had apparently chipped. "I don't know," she said finally. "I think I meant to, at the time, but looking back on it, I can't imagine how I even considered such a thing. I loved him too much."

"That's how I feel about Alex," I said. "When I was away, I missed him so much. If I've lost him, I'll never forgive myself. Do you think he wants me back?"

I expected her to say, "Of course he does!" But she didn't. She sipped her tea some more, and then said, "I don't know,

Lark. I guess you'll have to talk to him. I guess you'll have to decide if you're going to make a habit of walking out on him."

"I'm not! I won't ever again!" I vowed so ardently that I dribbled lemonade down my chin and had to wipe it off with the sleeve of my sweatshirt.

"Well, talk to him," she suggested again.

"I will! I will!" I cried, leaping up from my seat and heading out the door. She hurried out after me, and as I revved up my blue Grand Am, she appeared at my elbow.

"Honey, I'm glad you're home. I was really worried."

"I'm glad I'm home, too. And I'm sorry you were so worried. And I'm sorry for everything!" I began to cry, and she did too. We hugged through the open window. "I love you," I said. I heard her say that she loved me, too, and when she released me, we waved goodbye and I pulled out. I was going to talk to Alex.

I unlocked the door with my key, pushed it open, and poked my head in.

"Alex?" I questioned timidly. "Are you awake? It's me."

I heard movement in the kitchen, and then Alex's startled face appeared in the doorway. His face lit up when he saw me, and he threw himself into my arms.

"You're home! Thank God you're home!" he cried, pulling me inside and closing the door. Then he pulled away to study me. "Are you here to stay? Where have you been? Did I do something to upset you? I'm so sorry! I love you! I missed you so much!"

"I love you, too! And I missed you, too! Don't apologize! I'm the one who's sorry! I ran out on you, and I had no right to do that, and I never will, ever ever again."

184

He pulled me close and held me tight. "Why did you go?" he asked plaintively. I took a deep breath and led us over to the couch. We sat. He grasped my hand, kissed it, and stared urgently into my eyes.

"I don't know. I was unhappy and I didn't know why. I felt like I was wasting my life, like I was just sitting back, watching it go by. But when I ran away from it, I missed it, and I wanted it back." As I said that, I recalled with a wry smile that Nick had been right after all, that day he bought me pizza at Luigi's.

"So you're home for good? You feel better about everything?"

"I feel wonderful! Being away from you made me realize how much I love you and need you and can't live without you."

"Really? You can't? Thank God! I realized I can't, either. Live without you, I mean. Except I already knew that!" he babbled deliriously. I snuggled up to him. He smelled even better than I remembered. Through his shirt, I could hear his heart thumping, and it was a comforting sound. We both sighed. "So . . . where have you been?" he asked.

"Oh, Alex, you wouldn't believe it!" I laughed. "My car got stolen but I got it back, and I met this amazing clock repairman, and spent the night with a rich psychotic named Brian . . ."

"What do you mean, 'spent the night,'" Alex interrupted, and his green eyes narrowed. I was surprised.

"What?"

"I said, What do you mean, *spent the night*! You *slept* with someone while you were away? You walked out on me and got right into someone else's bed? What the hell is *wrong* with you?"

"Alex!" I was so taken aback that I couldn't even defend

myself right away. I'd never seen him so angry. He was jealous! He was jealous because he thought I'd been attracted to someone else! "Alex, I didn't do anything with him! I just spent the night telling him about my father! Really!"

"Really?" Alex repeated, and on his face was the desperate need to believe me. I still couldn't get over it.

"Really. I'd never make love with anyone but you," I assured him, taking his hand and holding it tight. "Never ever ever."

"Oh." He let out a sigh of relief. "Jesus! Well, um, so what else did you do?"

"I slept on the porch of a radio station."

"You did? Why?"

"Well, because I thought they might know something about the new Food of the Gods album. And they did, Alex! They said it will be out soon!"

Alex's face lit up and he said, "Hey, that reminds me . . ." He leapt up from the couch and disappeared into the bedroom. Our bedroom! I thought excitedly, Tonight I'm going to sleep next to Alex, and tomorrow I'm going to wake up with him!

When he re-emerged, I saw he was holding a package.

"I bought you a present. Today was your birthday, do you realize that?"

"It was?" I was startled. I'd forgotten all about it. "What did you buy me?"

As soon as Alex handed me the bag, I knew what it was without having to open it.

"Oh my God! The new Food album! It's out!" I shouted, tearing off the bag and scrutinizing it rapturously.

"Came out today," he said, pleased with my reaction. "I bought it this afternoon. The clerk had just opened the box, so I waited while he took out a copy and priced it for me."

"Wow! I can't believe this! This is so great! But, um, Alex . . ." I tore my gaze off my new album and studied him, puzzled. "What if I hadn't come home? You bought me a present . . . I mean, what would you have done if I hadn't come back?"

Alex sank onto the couch next to me.

"I don't know," he admitted. "I don't have the slightest idea what I would have done."

"You wouldn't have . . . you wouldn't have killed yourself . . . would you?"

He was startled.

"No," he said, "I would never do that."

I examined him. No, he would never do that. He was too strong. He was a lot like my mother. I thought of her, and a fresh wave of gratitude and affection and happiness spread through me. It was so nice to be home!

"Well, it's getting kind of late. Want to go to bed?" Alex broke into my musing somewhat awkwardly. He was giving me a hopeful smile, like he wanted to make love, but he would understand if I was too tired.

"More than anything in the world," I said, stroking his sweet face, and smiling at him significantly. Relieved, he took my hand and led me into our bedroom. I almost burst into tears when I saw our waterbed and my Audubon posters. "How could I have left this?" I murmured, snuggling up to him as tightly as I could without causing us to lose our balance.

"I don't know. Just don't ever again." Lowering me onto the bed, he kissed me passionately. Then his eyes met mine. "I mean it," he said. "While you were gone, I couldn't stand it. And I think . . . I mean, don't you think that . . . well, I know how you feel about marriage and everything, but don't you think it's time . . ."

His voice faded off. But I astonished him by saying, "I'd love to get married." In fact, he was so astonished, that he sat back and regarded me sternly, as if to say, "Don't joke about this," as if I was going to burst out laughing and tell him I was just kidding. But of course I wasn't, and I didn't. I just lay there, looking at him with eyes filled with love.

"Really?" he asked carefully.

"Really!"

"To me?"

"Of course to you!" I giggled and pulled him close again. Neither of us said anything for a little while. It was a very intense interlude. It felt good to have the heavy warmth of his body against mine. I thought, This is how everything was meant to be, this is where I belong.

We kissed. And while Alex was telling me how much he loved me and needed me, I realized I was exhausted, and dropped right off to sleep.

ABOUT THE AUTHOR . . .

The Dynamics of Flight is Robin L Stratton's tribute to what she calls "America at her finest."

"It was this idea of people helping people," she said, "that provided the theme of the book. I see it wherever I go. That's what I mean when I use the phrase, 'New Age.' A lot of people think I'm talking about pyramid power and Tarot cards and chanting incantations. But I'm referring to this trend where people are realizing their potential, not only as effective individuals, but as conscientious contributors to society. They're understanding that we all have to work together. It's a heightened consciousness. I like to think that the sense of isolation so prevalent in the 80s will disappear in the 90s. Being helpful and polite is definitely *in*."

Stratton's "Hip to be Kind" campaign continues in *New Witches*, the New Age sequel to *Raising The Pentagon*. Once again, Wendy and Scott discover there are powers beyond their ken, just waiting to be tapped.

PRAISE FOR

Raising The Pentagon

"A fantasy in the hard-boiled tone of a Raymond Chandler mystery of the 30s — with a subtle New Age plot . . . an offbeat prescription for hopelessness . . . should make a dandy movie."
The Brain/Mind Bulletin

"Thematic matter shifts widely from speculations about reincarnation and mysticism to sad and sometimes nearly bitter observations of contemporary society. Overall, the work is light and lively."
The Small Press Book Review

"Mystical *and* lighthearted — a rare combination! A good seller, adds a touch of fun to our fiction section."
Gita Beth Bryant
Unicorn Books